Synthetic Biology: A Very Short Introduction

VERY SHORT INTRODUCTIONS are for anyone wanting a stimulating and accessible way into a new subject. They are written by experts, and have been translated into more than 45 different languages.

The series began in 1995, and now covers a wide variety of topics in every discipline. The VSI library currently contains over 550 volumes—a Very Short Introduction to everything from Psychology and Philosophy of Science to American History and Relativity—and continues to grow in every subject area.

Very Short Introductions available now:

Available soon:

For more information visit our website

www.oup.com/vsi/

Jamie A. Davies

SYNTHETIC BIOLOGY

A Very Short Introduction

OXFORD
UNIVERSITY PRESS

OXFORD
UNIVERSITY PRESS

Great Clarendon Street, Oxford, OX2 6DP,
United Kingdom

Oxford University Press is a department of the University of Oxford.
It furthers the University's objective of excellence in research, scholarship,
and education by publishing worldwide. Oxford is a registered trade mark of
Oxford University Press in the UK and in certain other countries

© Jamie A. Davies 2018

The moral rights of the author have been asserted

First edition published in 2018

All rights reserved. No part of this publication may be reproduced, stored in
a retrieval system, or transmitted, in any form or by any means, without the
prior permission in writing of Oxford University Press, or as expressly permitted
by law, by licence or under terms agreed with the appropriate reprographics
rights organization. Enquiries concerning reproduction outside the scope of the
above should be sent to the Rights Department, Oxford University Press, at the
address above

You must not circulate this work in any other form
and you must impose this same condition on any acquirer

Published in the United States of America by Oxford University Press
198 Madison Avenue, New York, NY 10016, United States of America

British Library Cataloguing in Publication Data
Data available

Library of Congress Control Number: 2018935710

ISBN 978-0-19-880349-2

Printed in Great Britain by
CPI Group (UK) Ltd, Croydon CR0 4YY

Links to third party websites are provided by Oxford in good faith and
for information only. Oxford disclaims any responsibility for the materials
contained in any third party website referenced in this work.

Contents

Acknowledgements

First, I would like to thank Latha Menon of OUP for encouraging me to write this book on the first place. I am very grateful to members of my laboratory, and the Edinburgh synthetic biology community in which it is embedded, for very helpful discussions. I am, as always, grateful to Katie for her support, patience, and advice.

List of illustrations

Chapter 1
Biology: from analysis to synthesis

Living in interesting times

Synthetic biology—the creation of new living systems by design—is a rapidly growing area of science and technology that is attracting attention well beyond the laboratory and is provoking vigorous public debate. It is seen by some economists, government ministers, and leaders of industry as having the potential to transform productivity: David Willetts, UK Minister for Universities and Science, declared 'Synthetic biology is one of the most promising areas of modern science, which is why we have identified it as one of the eight great British technologies of the future. Synthetic biology has the potential to drive economic growth.' Others view it more sceptically: Jonathan Kahn, a specialist in legal aspects of biotechnology, described synthetic biology as 'the latest in a long line of claims of grand promise...associated with successive major biotechnological undertakings...[that] have not, as yet, come anywhere near realizing the extravagant claims made by their initial promoters.' It is seen by some commentators as a possible solution to a range of environmental and energy problems: Renee Cho of the Columbia University's Earth Institute, hoped that 'Synbio innovations could potentially help solve the world's energy crisis...[and] restore the environment by cleaning up the water, soil, and air.' Others are far from convinced: Jim Thomas of the ETC action group on erosion, technology and concentration, warned that 'Synthetic biology is a high-risk profit-driven field, building organisms out of parts that are still poorly understood. We know that lab-created life-forms can escape…and that their use threatens existing natural biodiversity.' It is seen by hobbyists as an opportunity for tinkering, for fun or for potential profit, in community workshops and garden sheds. Some observers view this activity as very positive: in an editorial for *The Scientist* magazine, Todd Kiocken wrote 'Citizen scientists are dedicated to education, innovation, and problem solving, using a new model in the human spirit of curiosity and exploration.' Others see an urgent need for such tinkering to be regulated: George Church, a prominent geneticist, suggested that 'Everybody who practices synthetic biology should be licensed, including amateurs. Same as cars, right? You're an amateur car driver, you get a license.'

The arguments have been stoked by hyperbole on both sides. New technologies often elicit extreme reactions, especially when they are considered in isolation and when they are presented—falsely—as something completely novel disconnected from the rich web of traditional sciences from which they emerged. The purpose of this Very Short Introduction is to present an overview of synthetic biology in its context, with as much balance as possible. There is no intention either to promote it as a technology or to argue for its repression: rather, the aim is to describe and illustrate the scope of synthetic biology and to provide an indication of its present and potential points of interaction with society at large.

Synthetic biology has been defined in many ways for many contexts, but the most general definition works by dividing biology as a whole into analytic and synthetic branches. Analytic biology, almost the only biology for most of the history of science, is concerned with understanding how naturally evolved living things work. Synthetic biology, by contrast, is concerned with the creation of new living systems by deliberate design. This definition is independent of the techniques used. It does not, for example, require any element of genetic manipulation: indeed, the research on creating life, described later in this book, has little to do with genes. Defined this way, synthetic biology ranges from the modification of existing organisms to do entirely new things, which is now routine at least at a small scale, to the as-yet unrealized creation of a living organism from non-living components. The subject is broad partly because of the way it is defined, and partly because it has two distinct and independent historical roots, one intellectually driven and running deep into 19th-century natural philosophy, and the other more practically orientated and emerging from late 20th-century biotechnology.

The first root of synthetic biology

One of the deepest biological questions asked by philosophers and scientists is whether life can be explained entirely by the natural laws of physics and chemistry. The 19th and early 20th centuries saw vigorous debate between materialists, who viewed life as exquisitely organized chemistry and physics, and vitalists, who held that living systems required something extra—an *élan vital* or *vis vitalis*. Though often nowadays dismissed as irrational or dogmatic, vitalists of the time used hard scientific evidence every bit as much as materialists did. One of the most famous experiments of 19th-century biology was that of Pasteur, who observed that sterilized broth would remain sterile if sealed but, if contaminated with a tiny number of microorganisms, would support the production of vastly more. The multiplication of the introduced microorganisms proved that the broth contained all the raw materials for making new living cells, but the necessity for 'seeding' it with a few living organisms showed that mere presence of the raw materials was not enough for life to emerge: something else was needed, something that could be provided only by the already-living. To vitalists this missing ingredient was the *élan vital*. To materialists, it was some function of organization that allowed a cell to produce copies of itself, organization that was missing in the soup of simple chemical ingredients. Both explanations fitted the data, and to take either position was a matter more of faith than of scientific proof.

Two very different approaches have been taken towards resolving the question of vitalism—analytical and synthetic. The analytical agenda aimed to gain a full mechanistic, physico-chemical understanding of how living things work. This approach was in any case the focus of much of mainstream biology—for reasons of scientific curiosity and because analysis of living processes was important to many practical problems in medicine and agriculture. Highlights of the analytical work of the last two centuries include Mendel publishing his theory of genetics in the 1850s; Friedrich Miescher discovering DNA in the 1860s; Theodor Boveri describing the chromosome-duplicating and -sharing processes involved in cell division in the 1870s; Theodor Boveri and Walter Sutton each demonstrating, in 1902, that specific genes are associated with specific chromosomes; Oswald Avery proving in 1944 that genes could be identified with the chromosomes' DNA; and James Watson and Francis Crick proposing in 1953 that DNA has a double-helical structure that would allow it to act as a template for its own copying. In the past few decades, vast numbers of researchers have determined how genes direct the synthesis of proteins; how some proteins in their turn control the activities of genes or metabolic reactions; and how the molecular machinery of the cell that separates chromosomes and then divides a cell to produce two daughter cells actually works.

This analytic work has given materialists a far greater ability to explain the physico-chemical basis of many aspects of cellular behaviour. It has not in itself, though, given them any way of disproving vitalism except by induction. Proof by induction, which is not really 'proof' at all, is a bedrock of science. It works by assuming that a pattern that has been observed in a large number of particular cases must be universally true. Knowing that humans, dogs, cats, bats, elephants, and hundreds of other mammals have four-chambered hearts, we state confidently that having a four-chambered heart is a characteristic of being a mammal, even though we have not dissected, and have probably not even discovered, all mammalian species. Induction is ubiquitous in science, but it is dangerous. Humans, dogs, cats, bats, elephants, flies, tube worms, and many other animals use iron-containing haemoglobin to transport oxygen around their bodies, leading to the 'rule' that this is a universal mode of oxygen transport. Unfortunately for induction, horseshoe crabs turned out to use copper-containing haemocyanin instead. Examples like this remind us that the fact that many aspects of cellular life have been described in physico-chemical terms still cannot be taken as logical proof that there are no vitalistic exceptions to this rule. The analytic agenda can refute vitalism robustly only when every single aspect of life, including such aspects as consciousness, can be explained. We may be in for a long wait.

The alternative, synthetic approach to resolving the debate about vitalism aimed to meet the challenge posed by Louis Pasteur's experiments directly: if life could be made artificially from non-living components, then the need for an *élan vital* could be discounted and the materialist explanation would be proven. Synthetic chemical approaches had already made a valuable contribution in this direction. In 1828, Friedrich Wöhler had synthesized the molecule urea, hitherto known only in the context of living organisms, from inorganic precursors. Although apparently not done with the vitalism debate in mind, his synthesis united the chemistry of the living with that of the inorganic world and gave cheer to materialists. Extending the scope of creative work from synthetic chemistry to synthetic biology was the next logical step.

Obviously the production of a complete living cell would be a tall order, so attention focused at first on reproducing specific aspects of cellular behaviour using non-living systems. One of the first major works in this area, usually taken to be the foundation of synthetic biology, was Stéphane Leduc's 1912 book, *La Biologie*

Synthétique. In this book, Leduc set out an uncompromisingly materialist agenda, insisting that life was a purely physical phenomenon and that its organization and development occurred by harnessing the organizing power of physico-chemical forces alone. He called this view 'physicism', and presented it in opposition to 'mysticism'. In order to demonstrate that there was nothing mysterious about the events observed in living cells and organisms, he constructed entirely physical systems that behaved analogously to cells. He stated, 'when a phenomenon has been observed in a living organism, and one believes that one understands its physical mechanism, one should be able to reproduce this phenomenon on its own, outside the living organism'. In modern language, the systems he built were not living but life-mimicking, or 'biomimetic'. Leduc was by no means the first to try to synthesize biomimetic systems from non-living components. Moritz Traube, in particular, had in the 1860s produced vesicles bounded by semi-permeable membranes by dripping glue into tannic acid, or mixing potassium ferrocyanide and copper chloride, and these were analogous enough to cell membranes that they could be used to study the laws of osmosis that applied to real cells. Leduc went a lot further, and used elaborate systems of diffusing chemicals to produce extraordinary simulacra of complex, biomimetic patterns (Figure 1). In describing the behaviours of these systems, Leduc argued that they showed, in addition to realistic physical forms, nutrition (the systems 'eating' simple components to use in building their own structures), self-organization, growth, sensitivity to the environment, reproduction, and evolution. He also argued, explicitly, that studying biomimetic systems may shed light on the ultimate origin of life far back in the history of Earth. Leduc was read closely by the great Scottish biologist D'Arcy Thompson, who cited him many times in his 1917 book *On Growth and Form,* a book that remains in print and is widely read by embryologists.

The impact of the early synthetic biologists on the intellectual thought of their time was surprisingly wide, given the very technical nature of their work. The Marxist polemecist Friedrich Engels, for

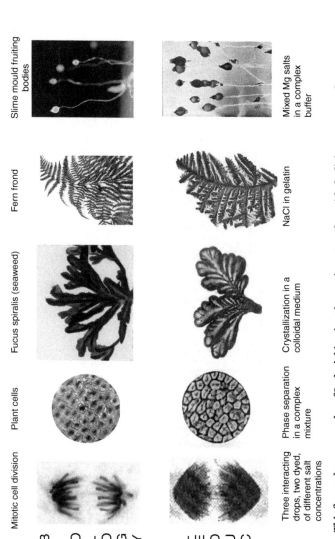

BIOLOGY

Mitotic cell division | Plant cells | Fucus spiralis (seaweed) | Fern frond | Slime mould fruiting bodies

LEDUC

Three interacting drops, two dyed, of different salt concentrations | Phase separation in a complex mixture | Crystallization in a colloidal medium | NaCl in gelatin | Mixed Mg salts in a complex buffer

1. This figure shows examples of Leduc's biomimetic creations, together with the living structures they were intended to simulate (Mg = magnesium; NaCl = sodium chloride (common salt)).

7

example, discussed Traube's vesicles in chapter 8 of his tract *Herr Eugen Dühring's Revolution in Science.* The Nobel Prize-winning novelist Thomas Mann devoted several pages in an early chapter of his allegorical work *Doktor Faustus* to a description of biomimetic systems studied around the turn of the 20th century by the father of the book's central character. Some of these systems were very similar to those of Leduc and included plant-like inorganic growths that grew in glass, some even being heliotropic (light seeking). One, the 'devouring drop' of oil in water, would, when offered a shellac-coated glass thread, deform to engulf it, strip the shellac off, and spit out the remains. Mann described the system so well in the novel that it is easy to recreate; I often set it up on an overhead projector to tease undergraduates. Mann uses the tension between vitalism and materialism to lay the foundations for later ambiguities between theological and rational explanations for morality, for genius, for madness, for cause, and for consequence.

Early synthetic biologists hoped to use biomimicry to persuade analytic biologists—embryologists in particular because embryology was vitalism's strongest hold—to seek the physical mechanisms that underpinned their science. In this they were largely misunderstood and even mocked: biomimicry was not seen to be relevant. At the same time, two competing new developments did offer the promise of 'mechanism', albeit of a different kind. The first was genetics, which in the first decade of the 20th century established associations between mutations in heritable genes and specific congenital changes in organisms such as fruit flies. The second was the discovery in the 1920s of embryonic induction—signalling by one part of an embryo to trigger a specific developmental event in another. These developments allowed embryologists to take a different kind of mechanistic approach; not as deeply mechanistic as physics would have been, but mechanistic in a 'black-box' sense of 'A causes B causes C', with the hope that one day the letters and the pathways of causation might later be reduced to physics and chemistry. In the second half of the 20th century, the era of molecular biology started to meet this hope. Genes were reduced to chemistry, and lines of clear causation, which were understood reasonably well at a molecular level, connected genes to proteins. The logic of biology came to be expressed not in the equations of physics but in chains of names, representing entities such as molecules or cells, and arrows representing processes.

The early flowering of synthetic biology before the First World War, grounded in biomimetic systems built almost exclusively for academic research, did not transform mainstream biology; its agenda, however, has never quite been forgotten. Small numbers of researchers, drawn from both chemistry and biology, have continued to work towards improved and more capable creating life-like systems, with the aim of one day making fully living systems from non-living constituents. Some scientists with a keen interest in the origin of life have worked on the problem of how complex organic molecules capable of making living systems could have arisen in the first place. A landmark experiment in this area was that of Harold Urey and Stanley Miller who, in the 1950s, showed that complex molecules, including amino acids, appeared spontaneously from simple precursors in simulations of the environment of primitive Earth. Other scientists worked on problems of organization; in the 1950s, Boris Pavlovoch Belousov described a chemical system that spontaneously produced complex patterns in space and time and, in the 1990s, Günter Wächtershäuser and his colleagues described complex metabolic cycles being organized on the surface of the mineral pyrite. Yet others have taken the presence of large organic molecules for granted and have worked on reproduction; from the 1980s, the laboratory of Pier Luisi has produced various systems in which simple membrane-bounded spheres feed on precursor molecules, grow, and reproduce. Most of this work has been done for one of two motives, both of which would have been familiar to Leduc. The first is to gain understanding about the origin of life. The other, still, is to meet Pasteur's challenge and to repudiate vitalism not by mere induction but by actual proof.

Foundations for biotechnology

The rise of genetics and molecular biology in the 20th century may have drawn biologists away from life-simulating and life-creating agendas, but they have been essential to the development of the other type of synthetic biology: the addition of designed new features to already-living organisms. The molecular revolution was critical to the development of modern synthetic biology in two ways, intellectual and practical. Intellectually, the molecular approach provided biologists with a mechanistic understanding of the inner workings of cells, which is critical to the challenge of designing new cellular systems to work alongside natural ones. These new cellular systems have a range of practical applications for the environment, for medicine, for engineering, and for chemistry, described later in this book.

One of the most important discoveries of 20th-century biology, made cumulatively by the work of thousands of researchers over many decades, is that almost all structural, chemical, and behavioural aspects of living cells depend directly on the activities of proteins. Proteins are made by the joining together (polymerization) of amino acids in an unbranched chain; most organisms have twenty different types of amino acid and each protein is composed of different amino acids joined together in an order specific to that protein. The amino acid chain folds into a three-dimensional structure that is controlled by its amino acid sequence and, in some cases, by interactions with existing proteins of other types. The three-dimensional structure and the chemical natures of the amino acids that compose it together determine the chemical activity of the protein. Organisms make many different proteins; humans make approximately 100,000 (depending on how one chooses to count them). Some proteins are inert but cooperate in making large-scale structures; the keratins in hair and fingernails are examples. Other proteins are chemically active and act as enzymes that catalyse biochemical reactions. Some chemical reactions are those of metabolism, for example the reactions that break food down to yield energy or that produce amino acids to make new proteins. Other reactions make chemical modifications to specific proteins, altering their activity. In this way one enzyme can control the activities of other enzymes or of structural proteins, and that enzyme may itself be subject to similar control. The tens of thousands of different proteins in a typical cell therefore form a vast regulatory network of mutual regulation and control. Because the activities of some proteins is altered by binding specific small molecules, such as sugars, metal ions, or waste products, the regulatory network of a cell is sensitive and responsive to its environment. The 'environment' can include other cells, and cells can communicate with one another using small molecules or secreted proteins. This communication can either be in the interests of both cells, for example in cooperating to build a body, or can be 'unintended' by one cell but useful to the other, as when an amoeba detects secretions from bacterial cells and moves to eat them.

Given the central position of proteins in running the chemical and regulatory processes of life, production of proteins, both natural and designed, has been and continues to be of great interest to biotechnologists in industry and medicine. Direct synthesis of proteins by chemical means is extremely difficult and only recently has it been possible to make anything more than very the smallest and simplest. Fortunately, the cell itself provides a solution. Synthesis of proteins in living cells is controlled by a simpler polymer, mRNA, a nucleic acid which consists of a long sequence of chemical subunits ('bases'). There are four types of base, abbreviated to A, C, G, and U, and their order is specific to a specific mRNA. The protein-synthesizing machinery of the cell binds to the mRNA and travels along it, reading the order of bases in sets of three and choosing which new amino acids to add to the growing protein chain according to these base triplets. The rules that translate the base sequence into the amino acid sequence form 'the genetic code'. The sequence of bases in each mRNA is itself determined in a 1:1 fashion by the sequence of bases in a broadly similar nucleic acid, DNA. At particular places along the very long strands of DNA in the cell are sequences of DNA bases that can act as docking sites for an enzyme complex that can make mRNA in a process called transcription. If the right combination of protein factors is present, the enzyme complex assembles on the DNA and starts to move along it, transcribing the sequence of bases in the DNA into the sequence of bases in a new RNA and continuing until it meets DNA sequences that cause it to stop. The stretch of DNA that can be transcribed into RNA is the physical gene. The RNA so formed will go on, perhaps after some processing steps, to be translated into a protein. In some cases, the protein made will be one of those important in activating other genes, so that genes are linked into a network of mutual regulation by proteins (and sometimes also by other biomolecules that are outside the scope of this introduction). To sum up, information held in DNA is copied into an RNA intermediate and used to direct the synthesis of a protein (Figure 2).

The discovery that cells use DNA as an information store to direct the production of proteins was critical to the development of modern biotechnology. Cells look after their DNA as a permanent archive and they reproduce it faithfully when they multiply, so that daughter cells have a full copy of the archive. The idea of synthesizing DNA and having cells use it to make a protein, rather than making the protein directly, makes a vast difference to the feasibility and economics of the biotechnology enterprise. In his book *The Black Swan*, Nassim Nicholas Taleb divided the economic world into two zones, Mediocristan and Extremistan. In the economics of Mediocristan, there is a close relationship between the work put into a project and the economic gain; a blacksmith makes money according to how many horses he shoes, a surgeon according to how many patients she treats, and so on. In Extremistan, the relationship between work and possible gain is broken because a difficult-to-make object can, once made, be copied with ease so there is no direct relationship between effort

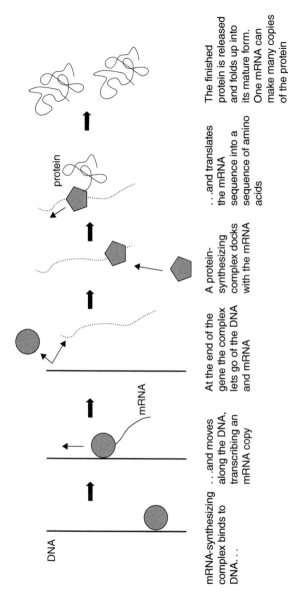

DNA

mRNA-synthesizing complex binds to DNA. . .

. . .and moves along the DNA, transcribing an mRNA copy

At the end of the gene the complex lets go of the DNA and mRNA

A protein-synthesizing complex docks with the mRNA

. . .and translates the mRNA sequence into a sequence of amino acids

The finished protein is released and folds up into its mature form. One mRNA can make many copies of the protein

protein

mRNA

2. Re-usable information held in DNA is used to produce proteins for making the structures of a cell and for running its metabolism.

in initial manufacture and the effort in making more units to sell. Novels, musical recordings, and software follow the economics of Extremistan. Direct manufacture of proteins would follow the economics of Mediocristan: making 100 grams of protein would need a hundred times as much manufacturing as making 1 gram. Indirect manufacture, via DNA, opens the gate to Extremistan: in principle, at least, once even one correct DNA molecule is made and put into a cell, it can be multiplied indefinitely along with the cells as long as the cell culture is fed, and all of the cells will make the desired protein.

Technologists' ability to construct DNA to order and to persuade cells to make proteins from it grew slowly, starting in the 1970s before it was even possible to read DNA sequences. Engineering of DNA began in two ways, one by direct synthesis of a gene and the other by cutting and recombining pieces of natural bacterial DNA to create novel combinations of genes. Yamada's laboratory was the first to synthesize a gene, in 1970, when they made an artificial copy of a gene from yeast. Paul Berg's laboratory published a technique for cutting and rearranging natural DNA in the same year, and in 1973 Stanley Cohen's laboratory succeeded in introducing this recombinant DNA into living bacterial cells that made a working protein from the introduced gene. The year 1977 saw a critical development in DNA technology with the introduction of two different methods for reading the sequence of bases along a DNA molecule, one from Walter Gilbert and the other from Fred Sanger (who had also been the first person to read a protein, in 1951). An ability to read DNA was crucial to the analysis of the connection between gene sequence and function; it is so much easier than reading proteins that protein sequences are typically determined nowadays by reading the corresponding gene rather than the protein itself. DNA sequences can now be read for fractions of a penny per base and can be custom-made for about 20 pence per base, meaning about a £100 for a typical gene; the process takes just hours. The technology has become cheap enough for large, complex projects to be undertaken.

The era of single-gene genetic engineering

Even before the advent of DNA sequencing, it was becoming obvious to researchers that DNA synthesis technology and recombinant DNA technology would create great biotechnological opportunities. As early as 1961, the pioneering French molecular biologists François Jacob and Jacques Monod speculated that gene-regulating elements could, in principle, be combined to produce custom systems for control of gene activity. In a far-sighted 1974 article, Wacław Szybalski considered this idea in the context of contemporary technical developments in DNA recombination and synthesis, and looked beyond analytic biology to the time when we would enter what he called 'the synthetic phase'. He wrote of the engineering of new genetic control systems, of adding these to existing genomes and even of constructing wholly artificial genomes, and explicitly referred to the emerging field as 'synthetic biology', a term he used again in a widely read 1978 article in the journal *Gene*. When writing this chapter, I asked Szybalski whether he coined the term 'synthetic biology' independently or whether he was making a deliberate link with Leduc's *biologie synthétique*: he replied that he had not known of the earlier work at the time. It is therefore reasonable to view the current, combined field as standing on two quite independent foundations.

Szybalski was prescient, but much work had to be done before his vision of synthetic biology would be a reality. The first practical applications for developments in molecular biology were much simpler and generally centred on the introduction of a single gene into a host organism to produce a desired product. Two early examples of medically important applications of genetic engineering were the production by bacteria of the human hormones somatostatin and insulin, in 1976 and 1979 respectively. Each of these hormones is a short sequence of amino acids, encoded by a gene in much the same way as larger proteins. Researchers transferred copies of the human gene into bacteria in the context of regulatory DNA sequences that would ensure that the gene would be active in these bacterial cells so that they would make made insulin: it is still made by an updated version of basically the same technique (details of how manipulations like this are actually done will be provided later in this book).

The first genetically engineered plant was produced in 1983, for purely experimental purposes, but by 1988 genes encoding human antibodies had been engineered into plants so that these precious proteins could be harvested in bulk. Later, genetic engineering came to be used for food crops as well as for medicinal proteins; in 1994, the tomato CGN-89564-2 ('Flavr Savr') became the first genetically engineered food to be licensed for human consumption. Unlike most genetically engineered organisms of the period, the tomato did not produce an extra protein but instead received a gene that interfered with the production of one of its natural proteins, an enzyme that is responsible for the transition of normal tomatoes from ripe to over-ripe to rotten. The engineered tomatoes had an extended shelf-life compared to normal tomatoes, as intended, but they were not a commercial success and were withdrawn in 1997. A similarly modified tomato was used to make tomato paste in the UK and the tins of paste, clearly labelled as coming from genetically modified tomatoes, initially sold so well that peak sales eclipsed those of the conventional alternative. In the late 1990s, though, a cultural change in the public perception of genetic modification caused sales to fall and the product to disappear from the shelves.

The 1980s also saw the first examples of genetically engineered mammals, made initially for research purposes. Developments at the end of the decade took genetic engineering beyond simple addition of genes and allowed geneticists to remove a gene of choice from a mouse's genome, or to modify the sequence of a gene in a very precise way. These techniques enabled them to create mice with genetic defects precisely analogous to those associated with human congenital disease syndromes. Pathologists were therefore able to explore the mechanisms of the disease and, in some cases, to test ideas for possible medical interventions. Translating the results of these studies back to human medicine has so far had mixed success because mice are not humans, and sometimes small differences between mouse and human physiology result in large differences in the effectiveness or safety of a treatment.

The early efforts of genetic engineering focused on the manipulation of just one gene connected with the final product, though a second small gene was often also added for identification and selection of the successfully engineered organisms. By the turn of the century, it had become more common for research-only genetic engineering projects to involve the introduction of multiple genes, and this tendency spread to commercial applications. An interesting example that shows the transition is the Golden Rice project, an attempt to address the serious global health problem of vitamin A deficiency that is common where rice is eaten as a dietary staple. The deficiency causes an estimated half a million cases of blindness and two million premature deaths annually. Humans can acquire vitamin A directly from animal sources or can make it from plant pigments such as beta-carotene. Natural rice does not contain much carotene, but, in principle, if rice could be engineered to make much greater amounts then the problem of vitamin A deficiency could be greatly reduced. Rice does contain an enzyme that can make beta-carotene from a precursor called lycopene (Figure 3), but the endosperm of rice, which is the only part that is eaten, contains little lycopene. Working with tomatoes in the 1990s, the biochemist Peter Bramley identified a bacterial enzyme, phytoene desaturase, that could make lycopene from a molecule called phytoene that occurs naturally in rice at modest levels. Engineering rice so that it expressed the phytoene desaturase gene in the endosperm created the first version of Golden Rice, so called because its enhanced production of the beta-carotene pigment gave it a golden colour. The yield of beta-carotene, though about five times higher than that of natural rice was, however, not thought high enough to eliminate vitamin A deficiency for people who ate only small amounts. Further

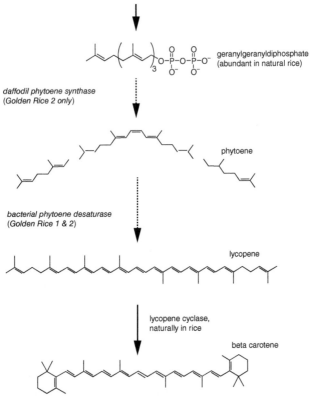

3. This figure shows the pathway engineered into rice to make it a richer source of beta-carotene, which human cells can use to make vitamin A. Reactions naturally available in rice endosperm are shown in solid arrows, and reactions naturally unavailable (or very inefficient), but performed by the newly added enzymes, are shown in broken arrows. The introduced genes are in italics.

analysis showed the disappointing yield to be due to the low amount of phytoene in the cells, suggesting that a second metabolic intervention might be beneficial. A daffodil enzyme, phytoene synthetase, was identified that could produce phytoene efficiently from a molecule called geranylgeranyldiphosphate, plentiful in the endosperm of rice. Introducing both the daffodil phytoene synthetase and the bacterial phytoene desaturase genes into rice created an efficient pathway for making lycopene, and thus beta-carotene (Figure 3). This version, 'Golden Rice 2', produces about a hundred times more beta-carotene than natural rice and has performed well as a beta-carotene source in trials on real human populations, although aspects of some of these trials connected to the principle of informed consent have attracted ethical criticism. The future of the project probably depends less on biochemical efficiency than on political, economic, ethical, and environmental arguments about the use of genetically modified food strains.

The second root of synthetic biology

The timing of the development of Golden Rice overlapped with the publication of the first constructions that went beyond what might be called 'traditional' genetic engineering and were described, both by their developers and by other commentators, as genuine 'synthetic biology'. Indeed, the Golden Rice project itself shows this transition. The first version of Golden Rice, with just one additional gene, seems rooted firmly in the era of genetic engineering. Golden Rice 2, with two new genes from different organisms that work together to make a significant modification to metabolism, works at a larger scale and, by making a multi-component system, has some of the character of synthetic biology. Though not as complex as the synthetic biological pathway used to alter the metabolism of yeast to make anti-malarial drugs (described in detail later), the construction of Golden Rice 2 clearly uses the same basic idea: if an organism cannot make something you want of it, identify enzymes from other organisms that can together create a metabolic path from a substance plentiful in your organism to a substance you want, and engineer your organism to have genes for those enzymes.

The year 2000 saw the publication of two landmark experiments that illustrated synthetic biological devices concerned with aspects other than metabolism. They did not directly address a commercial or societal need but were instead proof-of-concept constructions to stimulate thinking. James Collins's laboratory made a network of two genes, each of which would (via the protein it made) repress the activity of the other gene (Figure 4), and introduced it into bacterial host cells. Given that bacterial genes are active unless something is repressing them, the simple network had two states of stability: gene A on and gene B off, or gene A off and gene B on. The network had one extra feature: the proteins, which were borrowed from natural bacterial systems, could themselves be inhibited by either (A) a sugar-like drug or (B) an antibiotic. If the network were in the stable state in which A is on and B is off, but the sugar-like drug was added, the protein of gene A could no longer repress gene B, which would therefore switch on, making protein B, which would repress gene A and the system would now be in its alternative state. What is more, it would remain in that state even when the drug had gone. If someone added the antibiotic, protein B would no longer be able to repress gene A, and

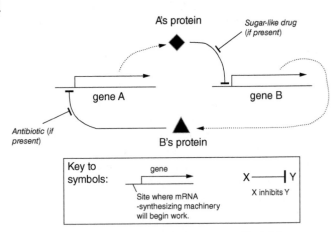

4. **The synthetic biological latch made by James Collins's laboratory changes its state on temporary exposure to one drug and then retains this state until exposed to the other drug.**

the state would revert to the A-active form. The network therefore acted as a memory, recording the condition (drug/antibiotic) that it experienced last. Indeed, its architecture is formally the same as that of a classical computer memory circuit, the set-reset ('SR') latch, although the computer version achieves in nano-seconds a state change that takes hours in the biological system.

The other dramatic experiment of 2000, from the laboratory of Michael Elowitz and Stanislas Liebler, used a very similar idea to generate an oscillator rather than a memory circuit. Again, genes encoded proteins that repressed the activity of other genes; however, this time there were three genes, A, B, and C (Figure 5). Gene A's protein repressed gene B, gene B's protein repressed gene C, and gene C's protein repressed gene A. The oscillating behaviour of the system depended on the fact that it takes a finite time (minutes) for an active gene to result in production of an active protein, and it takes a similar finite time for already-made proteins to degrade and disappear. Imagine a state in which gene C is active. Its protein will repress gene A, so none of A's protein is made. The lack of A's protein means that gene B is unrepressed,

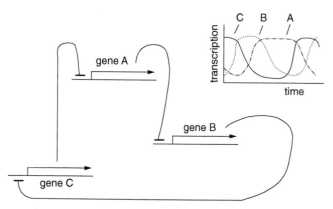

5. In the 'repressilator' of Elowitz and Liebler, each gene represses the next in the loop and therefore, indirectly and after delays, itself, causing the expression of each gene to oscillate.

so B's protein is duly made. B's protein shuts down gene C. Once the cell's remaining stock of C's protein has been degraded, there will be nothing to repress gene A, so A's protein will be made. This will shut down gene B. Once the cell's remaining stock of B's protein has been degraded, there will be nothing to repress gene C, so gene C will become active, C's protein will be made, and gene A will be shut down: we are back where we started. The system continues round and round, each protein following cycles of peaks and troughs in concentration. It is the analogue of a ring oscillator in electronics, and was named the 'repressilator' by its creators.

Synthetic genetic systems such as the latch and the repressilator were quickly followed by other oscillating and logic circuits, and these were connected to sensing systems so that the genetic logic would be responsive to a cell's environment. Systems like this have also been extended to multi-cellular systems, including mammals and plants. Many have been designed for specific purposes, in fields such as energy, the environment, and medicine. The danger of giving 'taster' summaries of these applications in an introduction such as this is that there is too little space for critical appraisal, and a consequent danger of an over-positive, misleading treatment: specific applications will therefore be described elsewhere in this book.

The latch and repressilator systems are described almost universally as 'synthetic biology'. What, then, is the difference between genetic engineering and synthetic biology? Different scientists offer different opinions, but, drawing answers together, we can at least sketch a rather fuzzy boundary. Scale is relevant, certainly, but more important is what might be called novelty of function. Expressing the gene for a human hormone in bacteria or the gene for a human antibody in plants is not 'synthetic biology' because the products of those genes have the same function that they would have in a human. The normal genes were moved to a different organism as a matter of manufacturing convenience, not because the desired product was intended to be novel. In a sense, the work is neither analytical nor synthetic but purely technical (this definition is not meant disparagingly: these projects can be very difficult and very important). Where, on the other hand, genes that never normally work together are combined to create a novel metabolic pathway, the outcome involves enough of a step away from evolved systems that most commentators would consider it to be 'synthetic'. Similarly, when even small numbers of genetic elements that are naturally present in an organism are rearranged to yield a novel function, such as performing computational logic, holding a memory, oscillating, or giving a warning of the presence of chemical or biological danger, the step away from normal biology makes these synthetic.

Applying engineering concepts to biology

A major (but not universal) school of thought in contemporary synthetic biology would add a few more characteristics: modularity, standardization, decoupling, and abstraction. All are related: they are borrowed from conventional engineering and can be illustrated most easily using examples from that discipline. Modularity is the idea that large constructions ought to be made by selection and connection of elementary building blocks. An example would be connection of simple modular components (resistors, capacitors, transistors, etc.), to make a circuit to perform a higher level function such as receiving a radio broadcast. Modularity can be hierarchical: an amplifier circuit made from a collection of components may itself be treated as a module that can be included in the design of radios, CD players, or answering machines. Modularity depends on standardization, so that designers can take for granted that the modules they use will have predictable, defined characteristics. Standardization can extend not just to reproducibility but also to the use of standard techniques for measuring parameters, to standard meanings of words, and to standards for connecting components (e.g. thread sizes of nuts and bolts). Decoupling is the idea that a complicated project can be broken down into a series of sub-projects that can be worked on in isolation. The building of a new radio, for example, can be separated into design, construction of components, assembly of components into circuits, construction of the outer case, etc., with those involved needing to be expert only in their own task. Abstraction is the related idea that systems can be described at a variety of levels of detail and that it should be possible for details at one level to be ignored and for the system to be treated as a 'black box' by someone working at a higher level. In the radio example, a component manufacturer has to pay attention to the detailed composition of, say, a capacitor; the designer of the tuning module can take the performance of the capacitor for granted but has to pay attention to how it and the other components are connected; the operator of the radio can take all of these things for granted and needs to know only how to use the buttons and knobs on the front panel. Like decoupling, abstraction is liberating because it avoids the need for everyone to be expert in every aspect of a complicated thing. It also allows designers to concentrate on high-level design without being overcome by detail. Importantly, under a hierarchy of abstraction, changes can take place at one level (e.g. a new material might be introduced for the capacitor) without higher levels being affected. How true this is in practice depends on how well the measurements taken to validate a component reflect all of its relevant properties.

From the mid-1990s, proposals were made for the construction of libraries of DNA modules that could be connected in many different ways for different purposes and some pilot studies were performed. In 2003, a major practical step towards the goal of having a library of standard parts for synthetic biology was made by Tom Knight and Drew Endy, when they launched the Registry of Standard Biological Parts and designed an international student competition to encourage its construction and use. These elements of the library were designed to be compatible with specific methods of assembly and include genes, DNA sequences that control transcription into mRNA, DNA sequences that ensure DNA is replicated, and adaptor DNA sequences that allow for flexibility in assembly. Some of the genes are themselves modular, so that designers can choose different subunits to give the resulting proteins different stabilities. At the time of writing, the registry has approximately 20,000 parts, most being for use in bacteria but some being for use in other organisms. It is fully open-access (parts.igem.org) and physical parts can be ordered from it by any academic laboratory.

In designing and naming their abstractions, enthusiasts for the use of these parts libraries have chosen language that emphasizes the 'engineering' nature of the approach and distances it from conventional biology. A clear example is the use of the word 'chassis' to describe a host organism. In the early days of electronics, before printed circuit boards had been developed, gramophones, wireless sets, and televisions were typically built around an inert metal box or plate—the chassis—on which sockets were mounted to carry the thermionic valves and other components that performed the function of the device. Referring to the host cell into which a synthetic biological device will be inserted as a 'chassis' encourages the engineer to regard it as equally inert and ignorable. This assumption is questionable, to say the least. Even the simplest bacterium is vastly more complicated than the synthetic device that will be placed in it; cells have evolved to react physiologically to changes in demands for energy and raw materials (such as those that the synthetic system might make), and they have evolved to fight invading genetic systems such as viruses. We still know so little about biology that it is impossible to be sure in advance that a protein made from a synthetic device will not interact directly with some component of the host cell in an unexpected way. In some projects, making these simplifying assumptions has worked well and in others it has resulted in failure. It is interesting that most of the successful projects used as case studies later in this book were not constructed using the approach of selecting standard parts from a library or, when they were, construction still involved a great deal of painstaking redesign and optimization. The value, or otherwise, of engineering-derived approaches and metaphors will be revisited in later chapters. Synthetic biology is a young science subject to many influences (Figure 6), and it is too early to be sure which approach will win out in the end.

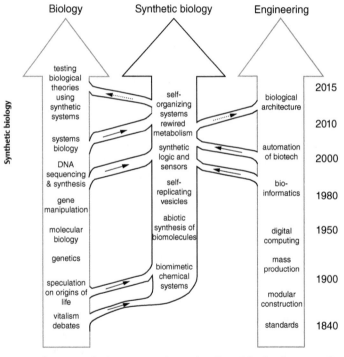

6. **The arrows depict an approximate time-line of the development of synthetic biology, indicating the main influences drawn from classical biology and classical engineering, and the contributions beginning to be made in the other direction.**

Applying biological concepts to engineering

One interesting effect of the synthetic biology enterprise, and the interdisciplinary science that led up to it, is that it has brought biologists and engineers together in a way that gave each significant exposure to the other discipline. Engineers, with varying degrees of arrogance or humility, have tried to impress on biologists the alleged advantages of the 'engineering' way of working. Biologists, again with varying arrogance or humility, have tried to impress on engineers the 'biological' way of constructing, maintaining, and evolving. Where the interactions happen to have been well-matched, some very interesting projects have arisen that attempt to apply living systems to problems that have traditionally belonged firmly in the inorganic worlds of silicon, stone, and steel. Some examples belong at the molecular level and include the use of molecules of DNA to carry encoded material, or the use of DNA to perform massively parallel computing tasks in realms such as decryption. Other examples operate at much larger scales and include work to create buildings that adapt to their environments or spontaneously heal after damage.

It is too early to say whether the influence of synthetic biology on engineering will be significant or will remain at the level of a few proof-of-concept studies. Indeed, it is really too early to make any prediction, with confidence, about the total impact of synthetic biology in any field at all. Chapters 3–7 of this book will present case studies of the early application of synthetic biological thinking to a range of global challenges, and will review both the science and the problems between design and commercial or humanitarian use, while the final Chapter 8 will consider current political, artistic, and cultural reactions to the technology. Hopefully, the information presented in this Very Short Introduction will allow an interested reader to reach their own informed, if provisional, opinion.

Chapter 2
How synthetic biology is done

Reading and writing the language of life

Synthetic biology, at least the kind that modifies existing cells, depends on technologies for reading and writing the DNA sequences of genes. Modification is almost always done at the genetic level because genes are heritable so have to be engineered only once. This makes unreliable engineering techniques tolerable; correctly modifying only one in a million cells is fine as long as it is possible to create an environment in which only that cell can survive and multiply to dominate the culture dish, leaving the failures long dead.

Reading ('sequencing') DNA is important for two main reasons: it is needed so that we can analyse natural genes and natural gene-controlling regions of DNA, and it is needed to verify the correctness of DNA we have tried to write. The most common method of reading DNA was invented in Fred Sanger's laboratory in 1977. Called 'chain termination', it uses a natural enzyme, DNA polymerase, that will, when faced with a single strand of DNA, make the other, complimentary, strand. DNA polymerase cannot begin work on completely single-stranded DNA, but it can work forwards from an already double-stranded section. To start the copying reaction, the experimenter provides a primer, a short length of single-stranded DNA that is complementary to (i.e. will pair up with) part of the single-stranded DNA to be read. The primer binds and creates a short double-stranded section; the DNA polymerase then goes to work to extend it (Figure 7). The need for a primer may seem to create a 'catch 22' situation in which a researcher has to know at least part of a sequence before reading can begin; however, in practice, a range of techniques is available for joining a piece of DNA with a known sequence to the end of a piece of DNA with unknown sequence, so that the primer can be designed to pair with the known sequence (Figure 8).

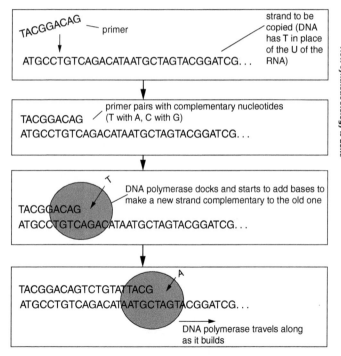

7. Primers allow DNA polymerase to make a new DNA strand complementary to an existing one.

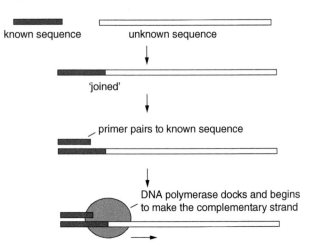

known sequence   unknown sequence

'joined'

primer pairs to known sequence

DNA polymerase docks and begins to make the complementary strand

8. Unknown sequences can be read by joining them to a piece of DNA with a known sequence, for which a primer is available. In this figure, the boxes represent single strands of DNA.

Being able to make a complementary strand from an existing strand is useful in many ways. Separating a newly formed pair of strands and repeating the process over and over again, for example, can allow researchers to make millions of copies from one piece of DNA in a process called the polymerase chain reaction (PCR), which will be mentioned again later in this book. Just copying, though, does not itself reveal anything about the original sequence. 'Reading' requires the deliberate poisoning of the reaction with traces of base units that have been modified in two ways: they incorporate a dye (each letter, A, T, G, C, has its own colour), and they are missing the chemical bonds that would be needed for the next nucleotide to be added to the chain. As the reaction proceeds, therefore, at every step there is a small chance that one of the special 'dye-terminators' will be incorporated instead of a natural base. This will both block any further elongation and give the stalled chain a distinctive colour according to which base was the last to be copied (Figure 9).

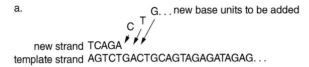

a.

G... new base units to be added

C T

new strand TCAGA

template strand AGTCTGACTGCAGTAGAGATAGAG...

b.

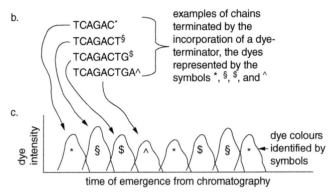

TCAGAC*
TCAGACT§
TCAGACTG$
TCAGACTGA^

examples of chains terminated by the incorporation of a dye-terminator, the dyes represented by the symbols *, §, $, and ^

c.

dye intensity

* § $ ^ * $ § *

dye colours identified by symbols

time of emergence from chromatography

9. Sanger sequencing operates by the low-probability incorporation of a chain-terminating, dye-containing base during production of a complementary strand.

The result is a mixture of differently coloured chains of different lengths. These are applied to a chromatography unit from which the molecules emerge in order of their size. A colour detector at the end records the sequence of colours as different length fragments come through one by one, and this sequence of colours will directly reflect the sequence of bases of the new strand, from which the sequence of the original can be worked out.

Nowadays, all of the processes involved in sequencing are heavily automated and one machine can run hundreds of reactions in parallel (in some systems, millions). This is useful because there is a limit to how long a piece of DNA can be sequenced before random errors create problems. For this reason, it is normal to cut long pieces of DNA into pieces before sequencing them. Computer algorithms have been developed that inspect the thousands of short lengths of sequence read, identify overlaps, and work out the only underlying long sequence that could generate all of those short sequences.

Writing in the language of DNA can be done in two broad ways. One is to make the entire desired piece of DNA by DNA synthesis, and the other is to cut useful pieces of natural or previously engineered DNA out of their host organisms and to join them together. This joining often includes the use of short pieces of purely synthetic DNA to provide sequence not available in nature, so there is usually still a need for some DNA synthesis. Construction of DNA from scratch is a chemical, rather than biological, process. It is difficult to string natural base units (nucleotides) together without the use of biological enzymes, so the most commonly used chemical process begins with modified nucleotides that have a reactive structure called phosphoramidite at the site where nucleotides will join. The chemistry involved in linking the nucleotides together risks making connections on the sides of the nucleotides as well as the intended head-to-tail chain. For this reason, the modified nucleotides also contain protective chemical 'caps' on potentially reactive locations to make them unreactive. These protective structures are removed (e.g. by exposure to acid) when the synthesis of the DNA is finished, leaving behind normal DNA.

The actual process of DNA synthesis (Figure 10) begins with one nucleotide being attached to a physical surface, and the chain is extended nucleotide-by-nucleotide in a series of 4-step cycles. Suppose we wish to make the sequence A-G-A-T...and that the first 'A' is already on the support, with an exposed 'end' (hydroxyl group) available for connecting to the next nucleotide (Figure 10a). In step 1 of cycle 1, the synthesis machine adds some G-phosphoramidate, 'activated' by exposure to a catalyst, to the reaction chamber. The phosphoramidate reacts quickly with the exposed phosphate on the 'A' and is therefore added to make the short chain A-G, connected with a peculiar, unnatural linkage

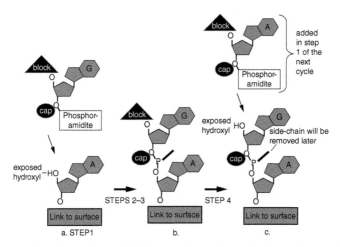

10. **Chemical synthesis of DNA: see main text for explanation of steps.**

with a side chain (Figure 10b). The incoming G-phosphoramidite does not itself have a free hydroxyl group because this covered by a blocking group to limit the risk of 'stutter' and of adding on more than one G. In step 2, a chemical process is used to cap any unreacted hydroxyl groups so that no more bases can be added to them. In step 3, an oxidation step makes the new linkage between the bases somewhat more natural, although it still carries a side-group: this step makes the chain more stable. In step 4, the blocking group of the newly added G is removed to expose its hydroxyl group, which will be available for the addition of a new nucleotide (Figure 10c). Cycle 2 will work the same way but will use A-phosphoramidite to extend the chain to be A-G-A, and cycle 3 will use T-phosphoramidite to yield A-G-A-T, and so on. At the end of the process, the DNA chains are released from their substrates and are chemically treated to remove blocking and capping groups, and to make the linkages shed their side-chains to become the phosphodiester links found in natural DNA. The chains are then purified to select only the ones of correct length, discarding any that missed a potential addition step and so were

capped (in step 2 of any of the cycles) and remained small.
If double-stranded pieces of DNA are needed, they can be made
either by synthesizing both strands and mixing them, or by
synthesizing a short primer, mixing it with one long strand, and
using DNA polymerase to make the rest of the missing strand
as in Figure 7.

This process of DNA synthesis outlined is rapid and has now
been automated so that many reactions (in some applications,
hundreds of thousands) can be done in parallel. It is not perfect,
however, and a small probability of error at each step means
that there is a limit to the length of DNA fragments that can
be synthesized reliably: this limit is currently about 1,000
nucleotides, the length of a small gene, with 300 nucleotides
the economically optimal length. Long, multi-gene constructs
are therefore made by joining short pieces together. Many
methods are available for this, but one of the most common is
Gibson assembly, named after its inventor, Dan Gibson.

In common with almost all methods for DNA manipulation,
Gibson assembly relies on the ability of two unpaired strands
of DNA with complementary sequences to pair up. When
someone plans to synthesize short lengths of DNA that will
be assembled to make a multi-gene construct, she does not make
sequences that follow perfectly on from one another but instead
arranges for there to be some overlap, so that the last thirty
nucleotides of one piece are identical to the first thirty of the next
(Figure 11: the 'thirty' is an average, not a rigid rule). Because
the sequence of nucleotides in the construct will not normally
have long repeated structures, the overlap sequences between
piece 1 and piece 2 will be completely different from the overlap
between piece 2 and piece 3, and so on.

The pieces of double-stranded DNA to be joined are mixed and
treated with an enzyme that has the property of chewing
asymmetrically a few bases into just one strand at each end of

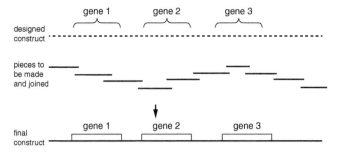

11. Constructing a long piece of DNA by designing a series of shorter pieces with overlapping ends that will be joined by Gibson assembly (Figure 12).

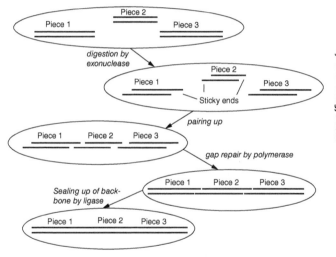

12. Gibson assembly works by chewing away short lengths of one strand to generate 'sticky ends' which stick together to generate the complete chain. A line represents single-stranded DNA.

double-stranded DNA, to leave 'sticky ends': single-stranded projections that will be able to bind a complementary sequence (Figure 12). If everything has been designed properly, the only complementary sequence present that can pair with the sticky end at the right-hand side of piece 1 will be the sticky end at the left-hand side of piece 2, and so on. When the pairing up ('annealing') is complete, the result will be a correctly ordered chain of DNA held together by nucleotide pairing, with some odd gaps in the backbone where the chewing back went too far. DNA polymerase can use information on the intact strand to fill in these gaps, and a third enzyme, DNA ligase, heals the DNA backbone to generate normal, intact double-stranded DNA. In practice, these reactions do not always run perfectly and the construct will be sequenced to check it. Errors will be repaired by various techniques for DNA editing, one example of which is the CRISPR reaction described later.

Designing the constructs to be built

Methods to read and write DNA are the critical enabling technologies for synthetic biology, just as reading and printing are the critical enablers for the practice of storing knowledge in libraries of books. In neither case does the technology itself help with the creative task of deciding what is to be written. Methods for making genetic systems lie on a continuum between two extremes. One is epitomized by natural evolution, and proceeds by successive rounds of random modification followed by selection; the other relies on rational design of every element from first principles. Between them are many hybrid ways of working. Typically, each element of a genetic system goes through the 'engineering cycle' (Figure 13). Given that genetic systems can, once built, be reproduced trivially as their host cells reproduce, most of the cost is in the design and testing phases rather than in final production.

As in other areas of engineering, the process of design begins with the setting out of a clear objective and, with that in mind, agreement on measures of success and of the specifications that the finished device must meet. The next stage of design usually

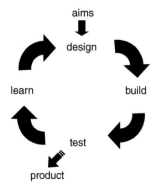

13. The engineering cycle, common to many fields of engineering including synthetic biology.

focuses on imagining a mechanism that could meet the specifications in terms of a set of connected modules, with the internal details of each module being left for later. Designing at this level is similar to electronics engineers sketching a circuit in terms of high-level modules such as amplifiers or filters, or to computer programmers summarizing their ideas as flow-diagrams or pseudocode. It is useful because it allows many possible designs to be sketched and compared without a great deal of time having to be invested in fine details. The comparison of rival designs may involve computer simulation, which is much faster than biological construction, and the outputs of the simulation may be used as a guide to modify the design and run the simulation again until the model of the device behaves as expected. One particularly valuable feature of computer modelling is that it allows designers to simulate their system with slightly different numerical values for factors like the efficiency of an enzyme or the time delay in activating a gene. This can indicate how tolerant designs are to variation in these parameters: generally, designs that tolerate great variations are useful and those that will only work if everything is within very tight limits are avoided.

Once a team has arrived at a promising design, the inner workings of the modules are designed. Depending on the initial level of abstraction, design may involve breaking each module into sub-modules, perhaps repeating this cycle several times before reaching the final level of DNA sequence (Figure 14). This way of tackling design, which is by no means unique to synthetic biology, has the great advantage that once their specifications are agreed the inner workings of different modules can in principle be dealt with in isolation, perhaps by different people in different places. In practice this is almost but not quite true. When designing DNA, for example, it is generally wise to avoid the same long sequence of nucleotides appearing in more than one place in the device because that can encourage host cells to perform a genome-editing process called recombination that will modify or destroy the device. At some stage, therefore, the inner details of the modules do have to be considered together.

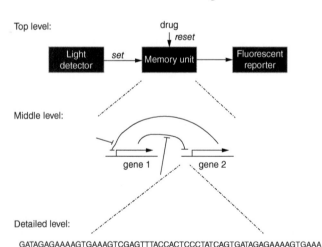

14. **Designing at different levels of abstraction: the top level has 'black boxes' with no internal detail, and more detail is added down the diagram.**

Taking modules and sub-modules 'off-the-shelf'

Once a module has been designed for one device and has been built and tested, it might in principle be useful for other devices as well. Again, this idea is common in engineering; the same amplifier module turns up in many consumer gadgets, the same graph-drawing library turns up in many software projects, and the same petrol engine turns up in a variety of lawn-mowers, pumps, and chainsaws. Community-minded synthetic biologists therefore contribute their modules, freely, to libraries such as the *Registry of Parts* (http://parts.igem.org/Main_Page) and any synthetic biologist can order them and use them. They can also modify them, perhaps sending the modification back as yet another contribution of the library.

The existence of libraries of freely available modules exerts an influence on design because it may be better, in terms of time and development money, to use readily available modules and to limit new design to components that connect them or perform some brand new function. The extent to which using modules really does save time depends on how well the characteristics of existing modules are known and how sensitive these are to the type of host cell and to environmental factors. Some sections of the synthetic biology community have begun to tackle this problem by proposing standards, initially for measurement and testing but perhaps ultimately for the minimum specifications of modules that deserve a particular name. The researchers involved in this have constructed a small number of modules that have been characterized very carefully, though it is still not possible to cover all imaginable conditions.

One often-expressed argument for the importance of ensuring highly predictable behaviour of individual components in a broad range of contexts is that the precision of a whole assembly can only be as good as that of its worst component. This plausible-sounding argument is, however, a fallacy. Good engineering practice uses closed-loop control, in which a measure of the output of a process is fed back to control the process itself. Feedback allows a high-precision system to be made with low-precision components. One of the first lessons most student electrical engineers learn is that there are two basic ways of making an amplifier with a known gain of, say, ten. One uses no feedback and relies on having transistors with precisely known characteristics: these are expensive. The other design feeds back a sample of the output to the circuit in such a way that the transistor 'sees' just enough of the input signal to make an output ten times greater: this circuit can use cheap transistors that have specifications with huge margins of error. The only components in such circuits that need to be predictable are the very few that are involved in the feedback loop itself, and even here it is common to include an adjustable component so that an engineer can make a final compensation for differences between the ideal and the real. As in electrical engineering, so in synthetic biology: intelligently designed closed-loop control can make the system as a whole very tolerant of variable performance at the component level.

Bearing all of this in mind, the balance of advantages and disadvantages in using libraries of standard parts for synthetic biological projects depends very much on the overall aim. For the short-term projects characteristic of student projects, libraries are very useful. They facilitate rapid design and assembly of devices that work well enough. For major, long-term projects that are intended to produce the best possible device for a real-world application, it may be much better to invest time in designing everything for the precise intended purpose and including existing modules only if they really do happen to be exactly right. Of the thirty-two projects described in the following chapters, only six made extensive use of synthetic biology libraries. This should be borne in mind when reading claims that libraries of standard components are at the centre of the synthetic biological enterprise.

The production of optimal systems may sacrifice the idea of rational design. We do not yet understand enough about matters such as enzyme structure to be able to design the gene encoding the best possible enzyme from first principles, so it is common for synthetic biologists to make a large range of similar versions of a gene and pick the best. Sometimes this involves very laborious sequential quantitative testing, but often it is possible to place the synthetic gene in the context of other modules that will allow cells to thrive only in proportion to how well the protein encoded by that gene works. Cells carrying the best versions therefore come to dominate the culture. There may even be subsequent rounds of mutation of that initial winner, to see if any small changes make it even better. This way of working, which mimics natural evolution, is not 'irrational' in the sense of foolish, but it departs from ideas of 'rational design' that assume a linear flow from knowledge to product.

Loading the synthetic system into a living cell

Once a synthetic construct has been made it must be loaded into a host cell. This is often done by building it into a piece of DNA that can inhabit a cell, at least temporarily, separately from the cell's own genome. The choice depends on the type of cell and the size of the construct. Bacterial cells naturally harbour plasmids: circular pieces of DNA a few thousand nucleotides long that contain special sequences to ensure that they are replicated by the host cell. Though this property makes plasmids seem parasitic, they often carry genes useful to bacteria, for example giving them resistance to antibiotics in clinical use. Plasmids can carry a few thousand nucleotides of engineered DNA: usually this will include a gene encoding resistance to a laboratory antibiotic. Plasmids are usually placed into bacterial cells by a somewhat crude process in which bacterial cell walls and membranes are damaged to make holes, and the cells are soaked in a solution of the plasmid so that at least some will take it up. Those will be the only ones that can survive the antibiotic when it is later applied.

Constructs too large for plasmids can be placed in bacteriophage (viruses that prey on bacteria), so that the synthetic DNA replaces most of the bacteriophage DNA. The result is a bacteriophage that can infect a bacterium and deposit its genome in the cell but cannot then proceed with a normal viral life-style. Larger still are bacterial artificial chromosomes (BACs), which can carry a synthetic cargo of few hundred thousand nucleotides. This is large enough for very sophisticated devices.

Yeast cells can harbour small plasmids and also yeast artificial chromosomes (YACs). YACs, like normal chromosomes, are linear rather than circular and contain sequences that allow them to replicate and move like normal chromosomes when yeast cells divide. They can carry around a million nucleotides of DNA: generally much more than is needed for contemporary synthetic devices but useful for major projects such as Yeast 2.0, to which we'll come later. Mammalian artificial chromosomes (MACs), including human artificial chromosomes (HACs), also exist. The use of MACs and HACs to carry synthetic devices separately from the host genome is relatively unusual, especially in mammalian cells: the artificial chromosomes are inconveniently large. More commonly, constructs are introduced into mammalian cells in bacterial plasmids or inside replication-incompetent mammalian viruses, with the intention of the construct becoming part of the host genome itself.

There are several ways of moving genetic material from a temporary carrier ('vector') into the genome, but one of the most powerful is gene editing by CRISPR. CRISPR is based on a system used by certain bacteria as a defence against bacteriophage, which are viruses that prey on bacteria. Such a bacterium holds on its chromosome a cluster of genes, each of which matches part of the DNA sequence of one particular bacteriophage. When these genes are transcribed, they produce pieces of RNA ('guide RNA') that combine with a DNA-cutting nuclease enzyme Cas9. If the RNA happens to bind matching DNA, for example that of an invading bacteriophage, the Cas9 enzyme is activated and cuts that DNA, breaking the genome of the virus and rendering it inactive. Introducing a system like this, that damages DNA, into a mammalian cell might not seem a promising technique; the power comes from systems already present in host cells that repair damaged DNA. One repair system, which is activated by the presence of single-strand breaks, is called template-directed repair and uses information from a 'template' DNA strand that has the same sequence as the broken strand in the regions each side of the break to rebuild the nucleotide sequence in the break. Normally, the 'template' DNA would be from the cell itself; in gene editing, the experimenter supplies the template. To perform the gene editing according to one currently popular method (Figure 15), the experimenter chooses a site in the genome where she would like her construct to be inserted, and flanks the construct with nucleotide sequences identical to the natural DNA of that site. Then she adds this DNA into a cell, together with Cas9 and guide RNA to target the Cas9 to the intended site. The result is a break that is repaired using the DNA supplied as a template, inserting the extra DNA into the genome. The method shown in Figure 15 works but is somewhat error-prone, and many cells end up not incorporating the device, meaning that some scheme is needed to detect and select the successes and discard the failures. More complex methods, using engineered forms of Cas9, have been developed to avoid this problem, but the core idea behind them remains the same. In general, experimenters prefer to insert their constructs in one of a few known 'safe landing sites': places in the genome that tend not to be altered, in terms of their ability to express genes, by the normal activity of the host cell. Again, there may be selection for antibiotic resistance to ensure that no cell benefits from expelling or shutting down the introduced module.

The CRISPR/Cas9 gene editing system shows in microcosm three important features of synthetic biology. The first is that its techniques depend heavily on basic research not initially directed

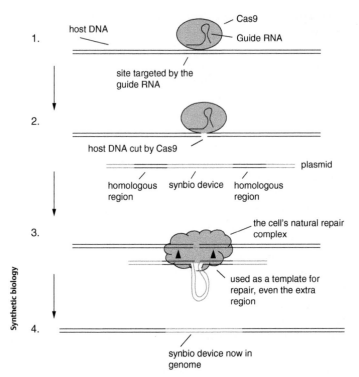

15. Using the CRISPR gene-editing system to insert a synthetic biological ('synbio') device into a host mammalian chromosome, for example in human cells in culture.

at any applications. CRISPR was identified in blue-skies exploration of the bacterial genome and only decades later did it become the basis of a useful tool; the same can be written for almost everything else that synthetic biologists use. The second feature is the 'hacker' mentality of using and adapting whatever is to hand, shown in this case by using Cas9 to co-opt a cell's own DNA repair systems to copy synthetic genes into the genome. The third is that the method is not 100 per cent reliable; unexpected and unhelpful things happen and there is always a need to select carefully the cells on which the technique worked and to discard the others.

Cross-talk and orthogonality

The stuff of synthetic biology is the stuff of life, and that creates a potential problem. Genes and control elements borrowed from evolved life will retain some of their original activities. If they are used in the cell types from which their natural forms originally came, they may well still interact with cellular control systems. Synthetic biologists may strive hard to limit the scope for interaction between synthetic and natural systems; to make them 'orthogonal', in the terminology synthetic biologists normally use to mean independent and non-interacting.

One reason to do this is safety. When synthetic biological systems are designed for use outside the laboratory, their designers consider ways to ensure that the devices can function only in their intended host and that they cannot survive in other organisms even if they are transferred to them by accident. One strategy for ensuring complete dependency on the intended host is to use unnatural amino acids to make the proteins encoded by the synthetic system. Normal organisms have a three-base genetic code that has sixty-four permutations. Cells have sixty-one transfer RNAs (tRNAs), each of which specifically recognizes one of these permutations and carries a specific amino acid so that, when the tRNA binds to its three-base target at a protein synthesizing complex (ribosome), the amino acid will be added to the growing amino acid chain of a protein (Figure 2). Three of the possible permutations of the three-base code have no corresponding tRNAs and act as 'stop' signals, terminating production of an amino acid chain. If all the instances of one of these stop signals in an organism are edited to become one of the other 'stop' signals, this now-unused three-base 'stop' code is free to be used to mean something else. The host's genome can be engineered to make a new tRNA that recognizes this former 'stop' code, and also to have enzymes that link this new tRNA to a twenty-first amino acid not used by normal life. None of the host's protein-coding genes will contain the repurposed three-base code so the host cell's metabolism will be normal except for the production of the new tRNA. If a synthetic biological device is designed to include the re-purposed three-base code, in its protein-coding genes, and the device is put into this special host cell, proteins will be made that include the twenty-first amino acid and are functional. If, however, the device finds itself in a normal cell of the same or a different species, there will be no new tRNA bearing the unnatural amino acid and its repurposed three-base code will simply be interpreted as 'stop': the protein will not be made and will not work, and the device will not function.

Orthogonality in synthetic biology is only a relative term. The core processes of all biology, whether synthetic or natural, make demands on the basic resources of a cell such as energy, raw materials, and enzymes that drive processes such as the transcription and replication of genes. Even when a synthetic system has been designed with great care to show no direct interactions of its genes and proteins with those of the host cell, competition for resources can still cause it to alter host cell behaviour. Enthusiasts for a high degree of orthogonality must therefore design systems that make only very modest demands on their host and alter neither its metabolic behaviour nor its fitness. The alternative, where safety considerations permit, is to forget about orthogonality, accept that interactions will happen, and design a combination of synthetic system, host cell, and environment that will perform the desired task efficiently as a unit.

Chapter 3
Synthetic biology and the environment

Twenty-first-century environmental challenges

The world faces a number of pressing environmental challenges: limited resources; limited space; loss of biodiversity; pollution of land, water, and air; and the impact of climate change thought to be driven partly by air pollution and by altered land use. Synthetic biology is being seen by many as a potentially powerful tool to be used in solving some of these problems in combination with other technical, social, and legislative advances. Examples of synthetic biology approaches to environmental protection include reduction of greenhouse gas production, more efficient agricultural land use, detection of pollution, and bio-remediation of polluted environments.

Biofuels to reduce greenhouse gas production

The burning of fossil fuels (coal, natural gas, and oil) currently accounts for more than 80 per cent of the world's primary energy consumption; globally, these sources generate an average power output of approximately 12 terawatts (TW; $1TW = 10^{12}$ Watts) with an additional 1TW being produced from terrestrial nuclear energy and a little over 1TW from extraterrestrial nuclear energy (as solar, wind, and hydroelectric power). Generating 12TW of power from fossil fuels is unsustainable: in the long term, the finite sources of these fuels will simply run out and, in the shorter term, rising atmospheric carbon dioxide (CO_2) from their burning may cause climate change that threatens even more economic damage than would shifting away from fossil fuels now. Direct solar, wind, and wave energy can be harvested by fixed power plants, but mobile vehicles require high-density, portable sources of power, and liquid fuels are particularly useful for this purpose. For this reason, there has been a strong interest in developing biofuels that are made sustainably from plants. Plants absorb CO_2 from the atmosphere and energy from the sun as they grow, and they release both again when the biofuel made from them is burnt: effectively the system is a carbon-neutral concentrator of solar energy. The problem with conventional biofuels, however, is that they are made from plants that grow on the same kind of land that can be used for agriculture; fuels and food therefore compete with one another for farmland and create pressure to convert more forests to farming, reducing biodiversity. Also, conventional plants are not very efficient at producing biofuels. Both problems could be solved by an organism that makes biofuels efficiently without needing farmland.

The most ancient photosynthetic organisms on Earth, and the cause of the first massive 'pollution' incident in the planet's living history—the release of oxygen into the atmosphere—are cyanobacteria, a form of bacteria that developed the trick of photosynthesis. Many can grow in simple liquid culture and proposals have been made for 'cyanobacterial farms' in areas of the world quite unsuitable for conventional agriculture; in these farms the cyanobacteria would live in sunlit tubes or tanks, turning CO_2 and water into biofuel. Photosynthesis, the process by which organisms capture the energy of sunlight to combine CO_2 and water to make organic molecules, is only a moderately efficient process in natural cyanobacteria mainly because certain enzymes involved (notably ribulose-1, 5-bisphosphate carboxylase/oxygenase and sedoheptulose-1, 7-bisphosphatase or 'RuBisCo'), are slow enough to create metabolic bottlenecks. Engineered alterations in the genes encoding these enzymes, including the use of genes from other species, have increased the efficiency of cyanobacterial photosynthesis. More dramatically, the laboratory of Bar-Even has been designing alternative, completely synthetic carbon-capturing pathways; for example, the malonyl-CoA-oxaloacetate-glyoxylate pathway, which can run two to three times faster than the normal mechanism of carbon capture. Overall increases in productivity have been disappointing, though, mainly because of the metabolic costs of building enzymes for the new pathway: improving on nature is less easy than some synthetic biologists assume.

Natural cyanobacteria keep most of the material that they produce by photosynthesis inside themselves for their own use, meaning that cells have to be harvested and destroyed if useful fuel is to be obtained. Synthetic biological techniques, coupled with cycles of mutation and selection, have been used to engineer *Synechocystis* cyanobacteria to excrete copious quantities of free fatty acids into the liquid that surrounds them: these can be used for biodiesel. Obviously this secretion represents a net cost to the cells, which are fragile and grow very slowly.

Another approach to biofuel production is to grow conventional plants and to use engineered microorganisms to turn their harvested tissues into biofuels efficiently, typically by fermenting them to ethanol. A great deal of biomass, including the waste biomass from food production (e.g. stalks and husks) is fibrous lignocellulose, which can be hydrolysed chemically to yield a mix of hexose and pentose sugars. In principle, these seem suitable for fermentation, but commonly used yeasts cannot cope with pentose sugars. Natural bacteria such as *Escherischia coli* (*E. coli*; a common laboratory species) can metabolize them to produce ethanol but only inefficiently, with unwanted acetate being produced as well. Synthetic biology techniques have been used to add enzymes (pyruvate decarboxylase and alcohol dehydrogenase II) from other organisms to *E. coli*, to add a new branch to the

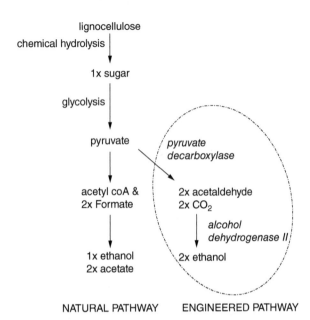

Synthetic biology

16. Use of synthetic biological techniques to introduce a new metabolic pathway into *E. coli* improves its production of ethanol from sugars obtained from fibrous plant wastes.

bacterium's metabolic pathways (Figure 16). With this in place, the bacterium produces mostly ethanol with hardly any acetate. Further optimization of the bacterium to increase its tolerance of ethanol poisoning has resulted in improved yields. Broadly similar strategies for introducing new genes to add metabolic pathways and protect cells from toxic effects have also been used to produce bacteria that make isobutanol or isopropanol instead of ethanol, these being easier to use in existing internal combustion engines.

At the moment, the most significant impediment to progress in biofuels is economic rather than technical. Relatively inexpensive fossil fuel hydrocarbons are cheaper than most current-generation biofuels and, except where taxation regimes penalize fossil fuel use or legislation requires the use of biofuels, there is little incentive for industry to invest in making what will appear to the consumer as an over-priced product.

More efficient food production

Photosynthesis, which uses the energy of light to combine CO_2 and water to form more complex molecules and release oxygen, has one great flaw: RuBisCo, the enzyme that should capture ('fix') CO_2 from the air, can fix O_2 instead in a process called photorespiration. In most plants, which use what is called the C3 pathway to perform photosynthesis, photorespiration wastes energy and reduces yield by up to a quarter. Some plants can use an alternative pathway (the Hatch-Slack, or C4 pathway) for photosynthesis: an effect of this pathway is to surround RuBisCo with CO_2, making it far more probable that it fixes CO_2 instead of O_2 and making photosynthesis more efficient. The C4 pathway has some energy costs, though, and as a generalization C4 plants have the advantage in warm, dry, low-soil-nitrogen conditions and C3 in cool, wet, nitrogen-rich ones. Some food plants (e.g. maize, sugar cane) use the C4 path, but most of the world's food crops (wheat, rice, etc.) do not; if a C4 version of these food crops could be built, they may produce higher yields and require less application of nitrogenous fertilizer. Synthetic biology offers one hope of achieving this, but the task will not be easy. Simply adding C4-type enzymes to a C3 plant is only part of the problem. In land plants, the C4 pathway involves cooperation between different specialized cell types that run different parts of the pathway, and plants rely on anatomical specializations such as leaf veins and enlarged cells for the CO_2-concentrating mechanism. Transferring all of this to a C3 plant, without interfering with its food properties, will not be easy. Current attempts therefore tend to focus on designing 'minimalist' versions of C4 pathways and trying to construct them in C3 hosts. This has not yet resulted in the desired metabolism; indeed, a recent (2016) review concluded that there is not yet even a realistic plan. A very different approach, not involving synthetic biology at all, is to use conventional mutation-selection techniques in environmental conditions that favour C4 plants, in a hope to drive evolution of C3 plants in this direction. This selection for extreme efficiency in adverse conditions does, however, run a serious risk of losing the useful food-producing properties of the plant, since making starch-rich seeds is arguably wasteful.

Another opportunity to make agriculture less environmentally damaging is to reduce its dependence on chemically synthesized nitrogen-containing fertilizers. Production of these is energy intensive (about 3 per cent of global energy production), and much of what is applied to fields ends up running off to pollute groundwater instead of being taken up by plants. Fertilizers are also expensive to transport, which is a problem to much of the developing world. Some plants (legumes, including peas and beans) can fix nitrogen from the air through an association with symbiotic bacteria: the bacteria produce nitrogen-containing ammonia within root nodules of the plant. Legumes are already used in crop rotation schemes to restore nitrogen levels in soil between seasons of growing other crops, but synthetic biology has raised the possibility of transferring nitrogen fixation to cereals and other bulk food plants. There are two basic strategies for doing this: engineering the biochemistry of nitrogen fixation directly into plants, or engineering root nodule formation into plants that do not yet have it. Neither is likely to be easy.

The bacterial nitrogen fixation pathway uses enzymes encoded by *nif* genes; these enzymes need metal-containing co-factors and are assembled in a complex process requiring the products of many genes. Furthermore, the pathway can run only in low oxygen environments, something that may be satisfied by placing the pathway in plant cells' oxygen-consuming mitochondria. At the time of writing, various proof-of-concept experiments have demonstrated components of the pathway in plants and even, most recently, the full array of 16 Nif proteins in a *Nicotiana* plant with some in mitochondria; however, a working, nitrogen-fixing pathway has yet to be achieved.

Root nodules are established through a signalling dialogue between plant and bacterium. Plant roots release flavonoids into the soil and these molecules activate *Rhizobium* bacteria to produce small modified sugar molecules called Nod factors; Nod factors cause nearby plant tissues to grow and produce nodules that the bacteria invade. The plant produces a membrane between the bacteria and its own tissues and the two species live together, the plant feeding the bacterium and the bacterium passing fixed nitrogen to the plant. This sequence is complicated, involving many genes for making and detecting signals, and also anatomical change in the plant. Many of the genes involved have been identified, but it is not clear how accurately they will reproduce the functions of nodule formation when placed into a new host.

Detection of pollution

Natural bacteria have evolved extremely sensitive receptor systems for detecting toxic compounds, often for the purpose of activating mechanisms to neutralize the toxicant before it harms them. These natural biosensors can be the basis of synthetic biological devices designed to detect and report pollution. An interesting example of such a device is an arsenic detector that originated in the work of an undergraduate team in the 2006 iGEM competition. The problem to be solved is that arsenic is present in the drinking water of up to a hundred million people, especially in countries such as Bangladesh where, in an attempt to avoid using surface water contaminated with disease, people sink tube wells unknowingly into arsenic-containing sediments. Only some wells are affected, so a simple sensor is needed that can detect arsenic in the region of 10 parts-per-billion (ppb) and that can be used in the field. The student team, from the University of Edinburgh, designed a device that combined existing components of the BioBricks library with new parts, including adapted genes

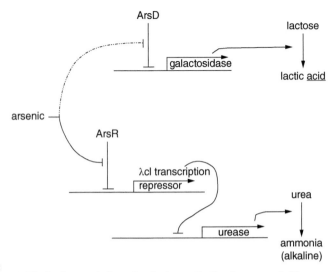

Synthetic biology

17. **The basic arsenic detecting device: activation is represented by arrows, inhibition by T-shapes, and the weak interaction crucial to the action of the system by a broken line. Arsenic switches the system in a concentration-dependent manner, from making alkaline urea to making lactic acid.**

(ArsD, ArsR) for a natural arsenic sensor, to translate arsenic concentration in water into a difference in pH that can be read with a standard pH indicator dye (Figure 17).

Since the initial development of the system, it has been improved by another iGEM team at the University of Cambridge, and The Wellcome Trust has funded development towards real-world application in collaboration with government agencies of Nepal. This has involved, among other things, packaging the system so that that is completely contained with little risk of synthetic biological organisms contaminating the environment, and so that they will not survive even if they do, and so that it is easy to use. It is not yet in use, though, for an interesting reason connected with regulation and ethics that is explored further in the final chapter of this book.

Bioremediation

One of the most easily recognized threats to the environment is industrial pollution, of air, water, and soils. In soils and water, in particular, even trace quantities of heavy metals can be problematic because of the way in which they concentrate in the food chain. Microorganisms absorb modest quantities from their surroundings and are eaten by predators that have no efficient methods for secreting the metal, so it accumulates. When these are in turn eaten by their predators, and so on, accumulation can reach dangerous levels. An infamous example of this, the 'Minimata disease' that affected over 2,000 Japanese people, arose from a factory releasing traces of methylmercury arising from a side-reaction of a mercury catalyst used in acetaldehyde manufacture. The mercury was taken up by plankton and accumulated through shellfish and fish to reach toxic levels in the humans (and cats) that ate them. Both industry and the wider public have a strong interest in ensuring that heavy metals are removed from waste; however, purely chemical methods are inefficient and/or very expensive when the pollutant is present in only trace quantities in large volumes of water. Given that the whole problem of heavy metal contamination arises through the ability of living organisms to concentrate them, adaptation of organisms to solve the problem is a promising idea.

Many bacteria have transporter channels that can take up metals such as nickel and cobalt (as ions), but the uptake is balanced by very efficient excretion pumps. Synthetic biological techniques have been applied to bacteria such as *E. coli* to enhance its ability to take up nickel and cobalt by adding channels from other bacteria, to eliminate its excretion pump system, and to add adhesion factors. The result is a bacterium ('Co/Ni Buster') that accumulates these metals strongly from dilute solutions in only a few minutes, and makes a sticky biofilm so that the metal-laden bacteria can be removed easily from the now-cleaned liquid. It is, though, only at research stage and has not been deployed in a real-world situation. Another example is a bacterium engineered to express the mammalian metal-binding protein metallothionein, which can sequester cadmium from soil and thereby increase the productivity of plants. Natural plants can also accumulate metals from soils and they have been used for bioremediation for some years: there is increasing interest in applying synthetic biological techniques to improve their ability to take up metals and to do so without killing themselves in the process. When the metals involved are valuable, their extraction from waste water may be a financially viable method of 'mining' (one possible application would be the recovery of precious metals from road drains contaminated with particles from catalytic converters).

Bioremediation can be used against a range of organic pollutants as well as metals. Examples of synthetic biological devices that have been shown to work at least in laboratory tests include bacteria engineered to remediate soil contaminated with explosives or organophosphate, atrazine, or pyrethroid insecticides, and waste water contaminated with medicines and caffeine.

For many pollutants, there is no simple enzyme system in any species that can be transferred to a host bacterium of interest. Here, there may be a need to design and produce entire new metabolic pathways by borrowing and adapting multiple enzymes and transporters from elsewhere—an approach that uses the true power of synthetic biology. This approach is supported by development of metabolism databases and accompanying artificial intelligence systems to suggest possible pathways from a starting chemical to a final one, in much the manner that a sat-nav plots a course from one place to another.

Barriers to commercial application

A recurring theme of this chapter has been the description of systems that have been shown to work in laboratories but that are not yet changing the world through real application. There are two reasons for this, one economic and one social. Biofuel production is inhibited chiefly by economics: crudely, as long as fossil fuels are cheaper in terms of direct purchase cost, there is no incentive to invest in systems such as algal biofuel production in deserts. Pollution detection and bioremediation applications are delayed not by economics (which generally favours them), but by highly restrictive regulations designed to ensure that genetically engineered organisms do not get loose and thrive in the general environment. Containment is difficult, especially in the context of remediating vast tracts of land or literal lakes of contaminated water. The applications described in this chapter may be technically rather simple examples of synthetic biology, but, paradoxically, they may be among the slowest applications to make a real difference to the way we live. The medical applications in the Chapter 4 are more complicated to build, but, because they are used only in highly controlled laboratory environments or operate in mammalian cells that cannot live independently in the environment, they are overtaking the simpler environmental applications and are seeing real-world application that is already saving lives.

Chapter 4
Synthetic biology and healthcare

Synthetic biology can be applied to medicine in several ways: it can be used to produce drugs and improve monitoring and diagnosis, and it is just starting to be used to modify human cells with properties designed to help patients. In a research context only, it is being applied to building new tissues. Medical applications make very high demands on safety for the individual patient; however, this is true of all medical developments and synthetic biological approaches are not particularly disfavoured in economic terms compared to conventional alternatives. For most applications, there is no risk of the engineered organisms 'running wild' in the general environment; risk is therefore restricted to specific patients who already have a problem that needs to be solved. In general, therefore, medicine is a field in which synthetic biology may make its earliest significant real-world contributions.

Drug synthesis through engineered metabolic pathways

Most drugs are small molecules that interact with, and alter the activity of, specific proteins in humans or in the microbes preying on them. Acetylsalicylic acid ('aspirin') is a familiar example: it inhibits the enzyme cyclooxygenase from synthesizing prostaglandin molecules that mediate inflammation and pain. Many important drugs are either natural products of plants and other organisms or, like aspirin, are synthetic chemicals based on them. Some natural products come from organisms that are difficult to cultivate. An example is artemisinin, a useful anti-malarial compound discovered by Tu Youyou that is a component of standard treatments for malaria worldwide, and is also useful against schistosomiasis. The drug comes from sweet wormwood (*Artemisia annua*); although the plant is cultivated in China, Vietnam, and East Africa, cultivation is difficult and prices fluctuate wildly, making it unaffordable by most of the people who suffer from a disease that kills more than a million people annually.

The problem of producing quantities of artemisinin cheaply was addressed by the pioneering synthetic biologist Jay Keasling, with funding from the Bill and Melinda Gates Foundation. Keasling's team modified the metabolism of brewers' yeast, which is easy to grow in fermentation tanks, by introducing enzymes from other organisms. Sweet wormwood produces artemisinin according to the metabolic pathway shown in Figure 18a: it uses sunlight to make sugar by photosynthesis, turns the sugar into acetyl-CoA, uses the mevalonate pathway to make this into the long-chain molecule farnesyl diphosphate, converts this to amorphadiene, which is oxidized to become dihydroartemisinic acid, which is turned into artemisinin by sunlight. Natural yeast has some, but not all, of the enzymes needed for these pathways (Figure 18b). Dae-Kyun Ro and colleagues, in Keasling's group, modified yeast by increasing the expression of enzymes in the mevalonate pathway to make farnesyl diphosphate more efficiently from Acetyl-CoA, by degrading a yeast pathway that would normally turn the farnesyl diphosphate into unwanted squalene, and introducing two enzymes from sweet wormwood to turn it into artemisinic acid instead (Figure 18c). Artemisinic acid is not artemisinin, but it can be converted to it by simple chemistry. Described like this, the metabolic engineering seems straightforward. It was not; the flows of molecules along different parts of the yeast's metabolism had to be balanced

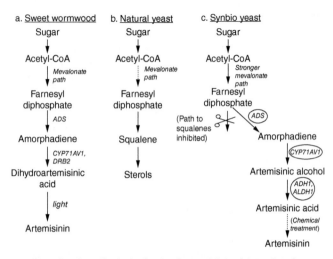

18. Natural and synthetic synthesis of artemisinin: (a) Depicts the pathway in sweet wormwood; (b) a pathway in normal yeast; (c) the synthetic biological pathway assembled in Yeast by Dae-Kyun Ro and colleagues. The enzymes in circles came from sweet wormwood.

very carefully for reasonable yields to be obtained, and the need for some enzymes was discovered only while debugging early versions. Since the construction of the metabolic pathway in 2006, it has been improved with further external support and is now being used commercially to produce artemisinin by Sanofi in Italy.

Artemisinic acid synthesis is synthetic biology's first great success story in terms of useful real-world application. It illustrates both the power of the approach and the difficulties in engineering new pathways to run efficiently, even when the work is being done by global leaders of the field. Broadly similar approaches, of reconfiguring the metabolism of easily grown microbes so that they make the unusual products of specific plants, are being taken to produce opiates such as morphine and codeine. Similarly, an enzyme from Ginseng has been transferred to yeast and optimized for use in this organism to make the anti-cancer drug ginsengoside Rh2 in a small-scale pilot experiment. Again, achieving this required some steps not initially expected, such as the need to inhibit yeast pathways that would otherwise degrade the wanted product.

Improving the immune system's armoury

A major thrust in cancer research aims to persuade the immune system to 'see' a tumour as a malevolent threat rather than to tolerate it. Tumour cells often express surface molecules that are unusual and these should in principle mark the tumour as 'foreign' and a target for immune attack. Unfortunately, the tumour environment often induces tolerance by the immune system rather than aggression. Recent years have seen the application of genetic engineering techniques to equip immune cells with synthetic recognition proteins to 'force' them to see the tumour and not to be restrained. Most examples centre on cytotoxic T cells, which the body uses to kill its own cells if they express 'foreign' proteins, as they would when a virus is growing in them.

All T cells express a recognition protein, the 'T cell receptor' (TCR), on their surfaces. The gene encoding the 'head' of the TCR, the part that actually recognizes a target, is mutated randomly as a feature of normal T cell development, so that different clones of T cells recognize different targets. Those that show strong recognition of normal body components are eliminated, as are those that recognize nothing at all. This elimination leaves the body with armies of T cells that recognize normal body proteins very weakly but that might recognize a specific viral protein strongly should it turn up. Normally, these T cells rest quietly. Another class of immune system cells, antigen-presenting cells, patrol the body continuously and collect samples of molecules from any sites that seem to be inflamed or injured. They present them to T cells in conjunction with a second, co-stimulatory, signal. The co-stimulatory signal alone does not activate the T cell, but, if the TCR of the T cell happens to recognize the molecule being presented, the signal from the TCR combines with the signal from the co-stimulation receptor to activate the T cell, leading it to attack any normal tissue cell that its TCR recognizes (Figure 19a).

Immunologists have worked out how to genetically engineer a synthetic TCR to recognize a specific, tumour-associated target molecule. This is done by replacing the normal TCR head by a synthetic module derived from an antibody that recognizes a tumour protein. The resulting protein is called a chimaeric antigen receptor (CAR; Figure 19b) and T cells carrying it are called CAR-T cells. If some of the patient's T cells are removed, engineered in this way and replaced, the TCR will recognize its intended target but may still not attack the tumour unless there is co-stimulation. Later generation CARs also incorporate internal domains that give both the TCR-type signal and the co-stimulation signal, even in the absence of any actual external co-stimulator (Figure 19c). In clinical trials, these cells have proved highly effective at attacking tumours. Unfortunately, they are often too effective, starting a 'cytokine storm' in which the immune and inflammatory systems become dangerously over-activated and cause life-threatening damage to the body as a whole. Clearly, before the therapy will be generally useful it needs to be placed under better control. One approach to this, typified by taken by Bellicum Pharmaceuticals' GoCAR-T design (Figure 19d), uses a 'plain' CAR without co-stimulation domains and adds a separate engineered co-stimulation receptor that is activated by a drug rather than by an antigen-presenting cell. The idea is that a physician can give a patient more or less of the drug to regulate how sensitive anti-tumour T cells are to activation on meeting a tumour cell. There are also many other advances of the CAR-T system under development.

In principle, the CAR-T idea can be used against targets other than tumours, including viruses and other parasites, but the

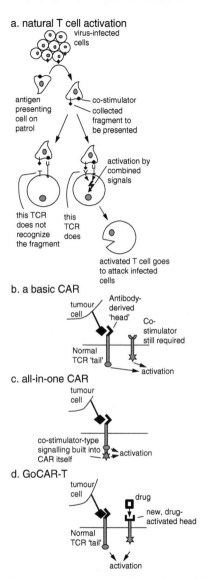

a. natural T cell activation

virus-infected cells

antigen presenting cell on patrol

co-stimulator

collected fragment to be presented

activation by combined signals

this TCR does not recognize the fragment

this TCR does

activated T cell goes to attack infected cells

b. a basic CAR

tumour cell

Antibody-derived 'head'

Co-stimulator still required

Normal TCR 'tail'

activation

c. all-in-one CAR

tumour cell

co-stimulator-type signalling built into CAR itself

activation

d. GoCAR-T

tumour cell

drug

new, drug-activated head

Normal TCR 'tail'

activation

19. CAR-T cells: (a) depicts natural T cell activation; (b) shows a basic CAR; (c) a CAR producing its own costimulation signals; and (d) the drug-regulatable, GoCAR-T system.

course of infection would need to be slow enough for T cell engineering to be done in time to help a patient. This is not yet feasible for rapidly growing viruses.

Making vaccines against new pandemics

Human history has been punctuated (and often shaped) by pandemics and we remain vulnerable to new pandemics, especially from viruses against which we still have rather few effective drugs. To give an idea of the risk, the 1918 pandemic caused by the H1N1 influenza virus killed 3–5 per cent of the global population, more than were killed by fighting in the First World War. Pandemics of H2N2 and H3N2 influenza virus each killed around a million people in the 1950s and 1960s. We do not have good drugs against influenza, but vaccination works well as long as it is against the right strain. The problem with outbreaks of new strains is that the viruses have to be isolated and cultured so that samples can be sent off to vaccine manufacturers. In the 2009 H1N1 outbreak, this took three months. A quite different approach that depends on synthetic biology, and which was tested by a consortium of labs in a 2013 H7N9 outbreak, begins with sequencing of the virus where it is first isolated. The sequence is then sent to vaccine producers directly, at the speed of e-mail: they then use it to engineer a benign lab virus to stimulate immunity to the real one and introduce it into cells for vaccine production. In a test, the process took not three months but a hundred hours.

A more daring technique anticipates virus strains that have not yet evolved but that have the potential to cause serious harm if they do. There has been a long-standing concern about the potential of H5N1 bird flu to mutate to something that can cause a pandemic in humans. Current H5N1 can pass between birds and from birds to mammals but not from mammal to mammal. An important question is whether it could mutate to spread between mammals and, if it does, what we can do about it. Ron Fouchier and colleagues in the Netherlands explored this question by engineering three specific mutations into the flu virus to make it more like mammalian versions. The virus was lethal to ferrets, but it still did not spread between them. Repeatedly infecting new ferrets with virus isolated from dead ones resulted in the isolation of a further-mutated strain that could now spread between ferrets. This is exactly what virologists were afraid could happen, but in the experiments H5N1 mutated to a lethal-in-mammals strain in the safe environment of a high-containment lab rather than in the wild. The random part of the experiment (giving virus from dead ferrets to live ones and waiting for mutations to arise) was repeated many times, and the resulting viruses were sequenced. Two mutations, in addition to the three initial designed ones, came up again and again, suggesting that they are important to this virus spreading between mammals. In principle, this approach to virology—deliberately making the very thing that we are scared might arise naturally—and then using it to make a vaccine to hold in readiness, would offer a very powerful insurance against a pandemic.

Synthetic biology in diagnosis

Diagnosis is a prerequisite to clinical treatment. The most important feature of a diagnostic test is precision, with speed also being critical for acute presentations such as possible infections. Where a bacterial infection is suspected, the classical diagnostic test is culture of a sample on agar plates followed by examination of the colony characteristics (colour, shape, smell, etc.): this may be combined with antibiotic-impregnated plates so that the sensitivity of the bacteria to a range of antibiotics can also be measured. The problem is that this takes a long time (typically overnight), during which the patient must be treated largely by intelligent guessing. Somewhat faster tests are available, based on antibodies or on PCR (Chapter 2), but these can test for only a very limited subset of the kinds of bacteria most often seen. One experimental application of synthetic biology in this area is the engineering of bacteriophage with reporter genes that cause infected bacteria to emit light. Different bacteriophage prey on different bacteria, and their specificity can be improved further by engineering. Mixtures of engineered bacteriophage carrying synthetic biological reporting devices can be applied to bacteria to identify them rapidly. Experimental systems working this way have identified the presence of *Yersinia pestis*, the cause of bubonic plague, in human serum within two hours.

Conventional diagnostic procedures usually require reasonable laboratory facilities, which is a problem in the developing world or on wilderness expeditions. James Collins and colleagues addressed this by producing a mixture of DNA and the enzymes required to transcribe genes and translate their messages into proteins, and freeze-drying it onto paper. Devices made this way are very cheap (pennies) and quick to develop because it is unnecessary to integrate them into cells for each round of testing. As a test, the team developed a simple synthetic biology circuit that normally repressed production of a fluorescent protein but would allow its expression in the presence of nucleic acid from a specific strain of Ebola virus. Their low-cost device proved able both to detect Ebola and to distinguish between strains found in different countries, even when they were present in tiny amounts.

In purely experimental, mouse-based systems, gut bacteria have been engineered to monitor aspects of their host's physiology and to report on it as they are expelled in faeces. If this idea were applied to humans too, it may be possible to construct a lavatory that reads signals from such bacteria and warns a user if anything is amiss.

Synthetic physiological systems

Many humans suffer from diseases that arise because some critical part of their body either never worked properly or no longer works properly due to injury, often of an autoimmune nature. An example is type I diabetes, which is conventionally controlled by injecting insulin that the body can no longer make in adequate quantities: this is inconvenient and it needs to be supported with very tight control of diet and exercise, because the dose injected is always an estimate of what will be needed rather than an amount made by the body in response to real-time measurements of need. Medical research has three broad responses to this problem. One is to build electronic pumps that release insulin in response to how much glucose is in the blood at any moment, and versions of these are already in clinical use. Another is to use stem cell technologies to try to produce or replace the missing cells, although this approach is problematic in a disease like type I diabetes that arises when the immune system destroys the critical cell type because the new ones may be destroyed too. The third response is to use synthetic biology to engineer a new system to perform the required task in another cell type. At the time of writing, synthetic pathways of this type have not been introduced into humans, but they are performing well in experimental animals.

Much of the synthetic biological work has been pioneered by Mingqi Xie and colleagues in Basel. To control type I diabetes in mice, the researchers engineered a human cell line with a combination of transport channels in the cell membrane that would, when the cell was in a low-glucose state, maintain a low internal concentration of potassium and calcium ions typical of any normal cell (Figure 20a). When the outside of the cell was exposed to high concentrations of glucose, however, the glucose flooded in through one of the channels the team had introduced, and was used by the cell as a fuel to generate the small energy-storage molecule, ATP. Rising ATP caused channels that normally remove potassium from the cell to close, altering the voltage across the cell membrane and opening channels that allowed calcium to flood in (Figure 20b). Calcium is often used by cells to control gene expression, and the team 'borrowed' components of calcium-activated gene expression from other

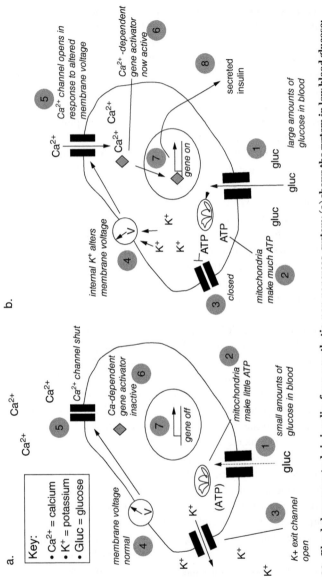

20. **Blood glucose control via insulin, from a synthetic, non-pancreas, system: (a) shows the system in low blood glucose; and (b) in high blood glucose, which causes production of insulin.**

a.

Key:
• Ca²⁺ = calcium
• K⁺ = potassium
• Gluc = glucose

b.

a.

small amounts of glucose in blood ①

mitochondria make little ATP ②

K⁺ exit channel open ③

membrane voltage normal ④

Ca²⁺ channel shut ⑤

Ca-dependent gene activator inactive ⑥

gene off ⑦

b.

large amounts of glucose in blood ①

mitochondria make much ATP ②

closed ③

internal K⁺ alters membrane voltage ④

Ca²⁺ channel opens in response to altered membrane voltage ⑤

Ca²⁺-dependent gene activator now active ⑥

gene on ⑦

secreted insulin ⑧

systems to link the rising calcium to the activation of genes encoding insulin. The result was a cell type that would respond to rising glucose levels by secreting insulin.

To test the clinical potential of their cells, the researchers introduced them into the bodies of mice that had type I diabetes. The result was that insulin levels were restored to normal and blood sugar levels, normally very high in these mice, fell to normal within three days and remained normal to the end of the study several weeks later.

A broadly similar system has been developed, again in mice, for the control of gout. Gout is an inflammatory arthritis caused by crystallization of uric acid in joints. Clinically, it is usually treated by drugs that increase uric acid secretion by the kidneys; there are also some newish drugs that reduce the amount of uric acid that is made in the first place. In principle, the problem could also be solved if some of the uric acid in the body fluids could be destroyed. Enzymes that destroy uric acid do exist: some fungi make urate oxidase, for example. The problem is that small amounts of uric acid are useful for protecting the body against oxidative stress, so engineering a body to produce too much urate oxidase would not be sensible. What is needed, as with diabetes, is closed-loop control so that cells sense how much uric acid is present and only destroy any excess.

Christian Kemmer and colleagues (in the same Basel lab) built such a system (Figure 21). At its core is a protein from the bacterium *Dienococcus radiodurans* that can bind either to uric acid, if it is present, or to a specific DNA 'operator' sequence upstream of a gene to block the activation of that gene. If a version of this protein is engineered into mammalian cells, together with a channel that allows uric acid to enter the cells in the first place, then it can be used to allow the presence of uric acid to control expression of fungal urate oxidase (Figure 21). If uric acid levels become too high, the cells make urate oxidase until uric acid levels

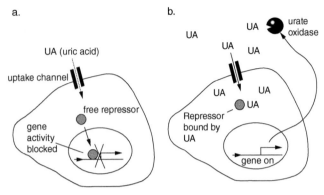

a.

b.

UA (uric acid)

uptake channel

free repressor

gene activity blocked

urate oxidase

UA

UA

UA   UA

UA   UA

UA

Repressor bound by UA

gene on

21. Use of bacterial, fungal, and mammalian components to make a synthetic, gout-controlling, closed-loop physiological system.

have fallen again. Having verified the action of their system in a culture dish, the team introduced the cells carrying the synthetic system into mice that were genetically prone to developing gout. The cells carrying the protein conferred considerable disease resistance on these mice.

There are many other diseases, some extremely common, that could in principle be controlled very well by similar closed-loop control: blood cholesterol, blood pressure, and the inflammatory activation associated with ageing are three important examples. The mouse experiments described earlier were all done with cells encapsulated within a device in the host, not free to roam, and the genome of the mouse itself was not engineered. The use of these systems in humans would not therefore require ethical acceptance of genetic engineering of human beings themselves, and adding encapsulated devices may not be more traumatic than the relatively common implantation of pacemakers, insulin pumps, and similar devices. There are still serious safety concerns, though, connected with rejection or accidental release of cells into the host. It may be that veterinary medicine will pioneer the use of engineered physiologies before they are used in humans.

Designer tissues

Tissue engineering is the construction of tissues to act as replacements for body structures that have either been damaged by traumatic injury or infection or that did not form properly in the first place because of a congenital disease. At present, clinically useful tissue engineering is restricted to structures that are relatively simple and homogenous at the cell level (cartilage, bone, skin, etc.) and it typically involves making an artificial biodegradable scaffold that can be populated with the correct cell type and placed in a patient: over time the cells replace the scaffold with biomaterials such as collagen that they make themselves, transforming the structure into one that is very close to natural. There is still an urgent need to be able to build more complex tissues such as kidneys and spinal cord, for treating patients in need of a transplant, and also perhaps to make un-natural, custom-designed tissues for repairing unusual bodies or for interfacing bodies with artificial limbs, eyes, or ears.

One of the most promising techniques for making complex tissues begins with stem cells (cells similar to those of an early embryo, which can make any part of the body). To make a tissue, say cardiac muscle, from these cells one 'walks them through' the sequence of signalling events that they would experience in a natural embryo, if they happened to be in the heart-forming group. The 'walking through' is done by applying hormones, drugs, etc., to the cells in a dish. If the process is done properly, the cells do indeed make the intended tissue type (in the case of the heart, one actually sees the cultured cells beating in their dish). The problem is that, while they organize themselves into a realistic tissue type, they do not make the shape of a whole organ because in life that is determined by interactions with other parts of the embryo that are not present in a dish of stem cells. A few tissue engineers operating at the forefront of the field are speculating on the possibility of using synthetic biology to engineer cells—not the stem cells but 'helper cells'—to provide the missing spatial information. The advantage of this approach, if it can be made to work, is that the stem cells themselves do not need to be engineered, reducing the risk of unintended changes being made to them: only the helpers, which can be engineered to kill themselves off at the end of the process, need carry synthetic biological devices.

The development of natural tissues, most of which happens in the womb, relies on a few basic cell behaviours that occur at different times, extents, and orders to build different structures. These include cell proliferation, cell suicide, cell migration, cell adhesion, cell fusion, sheet formation, and bending of a sheet (which includes making tubes from sheets). These actions are coordinated by a great deal of cell-to-cell signalling. As a first step to being able to make designed tissues, synthetic biologists have constructed a small library of devices that can be placed into human cells and, when activated, cause the cells to perform just one of these actions. The devices have been demonstrated to function as intended in very simple cell cultures, but they have not yet been connected to higher level control systems so that they make something useful. Whether this turns out to be easy or very difficult remains to be seen.

Chapter 5
Synthetic biology for engineering

While it may be expected that medicine and chemistry will have potential uses for synthetic biology, it is less obvious that the field might be relevant to the inorganic engineering of buildings, bridges, and computers. Nevertheless, the responsiveness of biological systems and the tiny scale of their constituent molecules do have the potential to solve some major problems (and perhaps to create some new ones). Again, in the interests of space, this chapter will illustrate a large field with just a few representative examples from three aspects of modern engineering: construction, computing, and communication.

Architecture and the built environment

More than half of the world's population now lives in cities, and the environment that surrounds most people, most of the time, is man-made. Constructing and reconstructing this environment accounts for at least 10 per cent of global carbon emissions, and heating, lighting, and cooling buildings account for even more. The contribution of buildings to human well-being goes beyond shelter, heat, and light, for there is a growing body of evidence that the nature of the built environment can exert a powerful influence on mental health and behaviour. Learning to build better is therefore an important priority.

Much of the construction in modern cities is based on concrete, often in association with steel. Concrete is a very useful building material, used since at least Roman times and accounting for the most widely used man-made material today. Transportable as raw ingredients, it can be produced on site, pumped as a liquid to where it is needed, and moulded into almost any shape. It does, however, suffer from a serious problem: its surface is inclined to crack, either in response to impact or through reactions between the alkaline cement and silica sand granules in the presence of water. In temperate and boreal regions, water penetrates cracks, freezes, expands, and makes the cracks worse. Hek Jonkers' team in the Netherlands developed a (non-synthetic) biological solution to this, in which *Bacillus* bacteria spores and dried nutrients are added to a concrete mix. As long as the concrete is intact, the spores remain dry and inactive. If it cracks and water soaks in, however, the spores activate and use the nutrient to grow and to secrete calcium carbonate, the cement-like substance that dominates sea shells and their fossilized remains. Thus cracks self-heal. The system works very well, at least in lab-type tests, and because it uses only normal harmless bacteria there are no great impediments to its being used in real buildings. It does have the problem, though, that it can confer self-healing only on concrete that has the spores in it at the time of its pouring.

The problem of applying the idea to existing concrete is being tackled by a team at the University of Newcastle. BacillaFilla, as they call their system, consists of a sprayable suspension containing spores of *Bacillus subtilis* bacteria carrying a synthetic biological construct, and nutrients (Figure 22). These spores are induced to germinate by alkaline conditions that are found in cracks in decaying cement. The synthetic module means that, once germinated, the bacteria become motile and at least some of them go deep into the crack; a quorum-sensing system borrowed from natural biology detects when large numbers of bacteria are present. When they are, the module causes the bacteria to become filamentous, to secrete a bacterial 'glue' so they lock

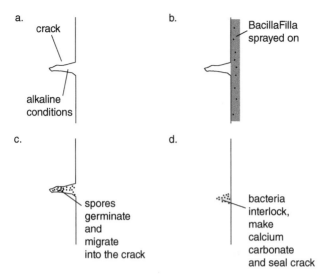

a.

crack

alkaline conditions

b.

BacillaFilla sprayed on

c.

spores germinate and migrate into the crack

d.

bacteria interlock, make calcium carbonate and seal crack

22. The idea for BacillaFilla, designed to be sprayed onto cracked concrete to repair it with calcium carbonate ('limestone') reinforced by cross-linked bacteria.

together, and to secrete calcium carbonate. The crack becomes filled with cement reinforced by interlocked bacteria. The system has been only part-built so far, so it is not yet clear how useful it will really be and whether its use in the general environment may be permitted by rules about release of genetically modified organisms. BacillaFilla does, though, illustrate how the tiny cellular agents of synthetic biology might have applications even at the scale of a tower block.

While practical synthetic biologists are making their first baby-steps in construction with designs such as BacillaFilla, dreamier types have recently been speculating about making parts of buildings living, particularly for energy management or maintaining the quality of internal air. Some go further, discussing possibilities for self-growing buildings. It is difficult to know what to make of this kind of very futuristic speculation. The futuristic predictions of the past (our all having jet-packs, flying cars, etc.) have seldom turned out to be correct; however, the images of futurists do often stimulate research that goes in entirely different directions.

Data storage in DNA

In the developed world, at least, the 21st century is characterized by a massive and growing infrastructure for storing and transmitting digital data. The amount of digital data in the world is estimated at around 2 Zettabytes (i.e. 2,000,000,000,000,000,000,000 bytes); it is expected to grow, and at least the more important data will need to be archived in some permanent form. Compared to existing electronic and optical storage media, DNA can store information at very high densities, about a tenth of a Zettabyte per gram. In principle, a copy of all the digital information in every computer in world, including additional information for indexing, could be stored in less than 50 grams of single-stranded DNA. This is about the same weight as a large coffee mug. In contrast to what might be expected of 'soft' biological systems, DNA is remarkably stable compared to tapes, discs, and CDs. About four-fifths of the genome of a woolly mammoth that has lain dead for over 10,000 years has turned out to be readable.

As digital text is encoded in strings of ones and zeros, it is an easy matter to render it into DNA by using the convention that either an A or a C = 0, and a T or a G = 1. In principle one could be more efficient in purely mathematical terms, but this coding allows design of DNA sequences that avoid long runs of G or C, which cause practical problems in making and reading DNA. George Church and his team used this idea to encode the 53,426-word text and eleven illustrations in Church's book *Regenesis: How Synthetic Biology Will Reinvent Nature and Ourselves,* in DNA. The text was divided into ninety-six-base data blocks, each flanked with a sequence start site and an index code, to make 159 base-pair modules. The flanking sequences allow the modules to be replicated, and also allow them to be sequenced for recovery of the information. When all of the sequences have been obtained, the index code allows the data sequences to be reassembled in the correct order. The group showed that they could indeed read the book back from the DNA. Other encoding schemes have been tried, with similar success.

The cost of DNA synthesis and reading, and the time required, makes DNA storage ridiculously expensive as an alternative to conventional hard drives for storing frequently accessed data. It may, however, come into its own for infrequently read but important archives designed to last forever. Conventional storage on magnetic media requires reading and rewriting of data every five to ten years (before the magnetism fades). It has been calculated that for really long-term storage, of five centuries or more, DNA-based methods would already be more cost-effective. DNA itself is immune from the problem of computer technologies becoming out of date, although the problem of accurately transmitting instructions for using these archives for readers millennia in the future will be a problem as language itself changes a great deal on that timescale.

Computing

Computer scientists divide problems up into classes, based (roughly speaking) on how quickly the difficulty in solving a problem rises with its size. Many problems, such as adding a list of numbers, are not hard: double the length of the list and you simply double the time taken to add. These are 'P' problems in the language of computer science. Others, though, can become exponentially harder as the size increases. Consider being given a simple set of numbers (e.g. −6, −4, 30, 45, 11, 9, −10) and having to find out if any subset of them add up to zero (as does, in this example, the subset −6, −4, −10, 11, and 9). Even the quickest way of doing this computationally works by taking possible subsets and trying them, which takes $2^n - n - 1$ tries where n is the number of numbers in the set. Thus, although answers can be verified trivially, the time taken to find them rises approximately to the power n. If solving the problem for a set of four numbers takes eleven seconds, doing so for sixteen numbers would take eighteen hours; for twenty numbers, twelve days; and for a hundred would take very much longer than the age of the universe. These 'NP' problems limit our ability to compute answers, because we cannot solve them unless they are really small.

Many NP problems (like the one just discussed) involve a brute-force checking of possibilities, which is slow if there is only one processor to work on them one at a time. Even machines that computer scientists call 'massively parallel', with thousands of processors, are defeated as problems get larger. In a test-tube, though, there can be trillions of separate DNA molecules randomly colliding with one another; if it were possible to encode a mathematical problem in the binding of DNA, this test-tube could race through staggering numbers of possible solutions in seconds.

In 1994, Leonard Adleman built a pioneering DNA-based computer that proved capable of solving an NP problem; it was not quicker than a silicon computer, but it did provide an important proof of concept. The Hamiltonian path problem chosen by Adleman begins with a set of dots joined by lines that can be traversed only in one direction (a 'directed graph' in mathematical language) and asks whether there exists a path by which one could navigate the graph from a defined entry dot to a defined exit dot and visit each dot exactly once (Figure 23a). To build a DNA-based computer to solve this problem, Adleman represented each dot of the graph by a specific twenty-base DNA sequence, the dot's 'label'. He also represented the lines between two dots as twenty-base sequences, these being constructed of the last ten bases of the dot label at the start of the line and the first ten bases of the dot label at the end of the line (Figure 23b). A line that can be traversed in two directions was coded as two separate one-directional lines. For the very start dot and the very end dot, the line was represented

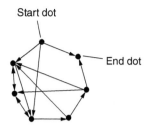

Start dot

End dot

b.

label of one dot

label of another dot

ACTGCCGTATCATTGCAACT

CTACATGAGCTGTTCAACTC

identical sections

CATTGCAACTCTACATGAGC

label of line joining those dots

c.

complementary dot labels

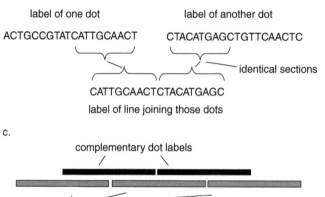

line labels held together by matching complementary dot labels

Synthetic biology for engineering

23. Computing with DNA: (a) depicts the Hamiltonian path problem; (b) shows the way that dots and lines are represented by single-stranded DNA; (c) shows how base-pairing creates long concatenations that represent paths through the system.

by the whole dot label. Finally, complementary copies were made of all the dot labels (that is, DNA sequences that could line up with the dot labels to make double-stranded DNA).

To run the computation, Adleman mixed all of the DNA sequences that represented lines with all of the complementary sequences to the dot labels. Even in his small tube, there were over ten trillion pieces of DNA randomly colliding and annealing to make double-stranded pieces where they could; with such a large number, all possible combinations would be tried many, many times. Where two lines came together at a dot, the complementary sequence to the dot label would stick the line labels together side-by-side. If a full path from beginning to end existed, then the path would be represented in a full-length piece of DNA representing all lines in their correct order, stuck together by the dot labels (Figure 23c). At the end of the reaction, pieces of DNA representing the start and end dot labels were used to prime a PCR DNA amplification (repeated replication) reaction, and the resulting pieces of DNA were separated according to their size using electrophoresis. The presence of a band corresponding to $20n$ bases, where n is the number of dots, showed that there was a path with the right number of steps, but it did not itself prove that each step was used just once. This was verified by testing that a single-stranded version of the DNA could bind every complementary dot label; the combination of a length that was the right number of steps and proof that each step was used at least once is enough to prove that each step must have been used exactly once.

Clearly, this method (which took about a week), is far less convenient than solving the problem on a desktop computer. DNA computers, at least as currently made, are custom-designed for a particular task rather than being general-purpose, and have error rates far in excess of those in electronic computers. In principle, however, a much larger Hamiltonian path could have been solved by the DNA method in approximately the same time and, for some specialized, large NP problems important enough to be worth the trouble, computing with DNA may be the only way to get an answer. There have been problems in the past for which it has been worth constructing a specialized machine at great expense: Tommy Flowers's encryption-breaking Colossus, at Bletchley Park, is a once-secret but now famous example. It presumably has equally specialized successors.

Cryptography

Probably for as long as information has been written in any medium, people have wished to control who can read it; as early as 1500 BCE, Mesopotamian craftsmen were encrypting text on clay tablets to protect their trade secrets and, 1,000 years later, the Greeks were using steganographical techniques to conceal the presence of a message. Both cryptography (message-encoding) and steganography (message-hiding) have been achieved for information held on DNA. There are many different schemes and the ones illustrated here have been chosen for being easy to explain in a short space.

A classic problem in cryptography is how one person, Alice, wants to send a message to her friend Bob without the eavesdropper Eve being able to read it. One way of sending images (pictures or images of text) was invented by Ashish Gehani and colleagues. It begins with two identical 'chips', divided into thousands of pixels, with each pixel carrying a different short strand of DNA (such 'gene chips' are commonplace and are made for micro-array analysis of gene expression). Alice and Bob each have one of this pair of chips, and the details of the chip do not have to be secret. They also each have a copy of a long piece of single-stranded DNA that does have to be kept secret. Within this long DNA are many pairs of short base sequences that use only the bases T, C, and G, each pair being separated from the next by a 'stall' site that consists of a short string of 'A's. The pairs of sequences each consist of one 'plaintext' sequence and one 'codetext' sequence, that are related to one another only by their position in that secret piece of DNA (Figure 24). Each codetext sequence is the same as one of the sequences on the chip. Alice also has a library of short pieces of DNA, each of which will bind to one specific plaintext sequence and prime a DNA replication reaction. She runs the DNA replication reactions with only bases C, G, and A available, so that each will stall when it meets the string of As (which would

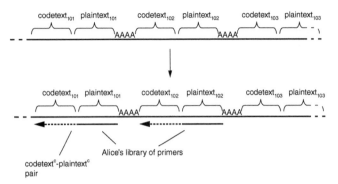

24. DNA encryption of images, part 1: Alice uses her library of primers to make codetext^c-plaintext^c pairs.

need a T to be added to make the complementary strand). The As therefore act as 'stall' signals. Furthermore, she uses special bases that are light-sensitive (and she performs her reactions in the dark to protect them). The result of adding the whole library to the long DNA, letting the copying happen, and then dissociating the copied DNA from the original is that Alice will have a collection of plaintext^c-codetext^c pairs, the codetext^c being photosensitive (the superscript c indicates a strand complementary, so able to bind to the original DNA).

Alice then adds her mixture of plaintext^c-codetext^c pairs to her chip, and the strands will hybridize to the complementary, immobile strands on the chip. She then projects her secret image onto the chip. In dark parts of the image, the pairs will survive. In the light parts, the light cleaves the DNA, liberating the plaintext strands (Figure 25). Alice collects these, and sends them to Bob. She then heats her chip to remove all the codetext^c strands, and destroys them.

When Bob receives the liberated plaintext^c strands from Alice, he uses them to prime DNA replication on his own copy of the secret long DNA, again using just C,G, and A, so that the stall

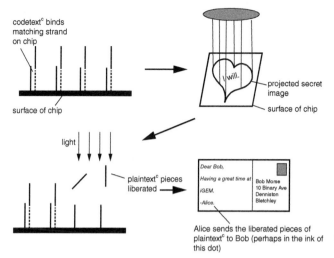

25. DNA encryption of images, part 2: Alice encrypts her image by applying her mix of codetext[c]-plaintext[c] to her chip, projecting her secret image, and collecting liberated plaintext[c].

signals work. He does not use light-sensitive bases but uses fluorescent ones instead. Because Bob only received from Alice the plaintext[c] strands that came from light areas of the image, his reaction will make only the plaintext[c]-codetext[c] pairs that correspond to the light parts of the image. He applies these to his chip, illuminates it with ultraviolet, and takes a photograph of the resulting fluorescence. The original image appears (Figure 26).

If Alice's consignment of plaintext DNAs message is intercepted by Eve, they will mean nothing. They will still mean nothing even if she has stolen a chip, because it is the codetext, not the plaintext, that matches the chip. As long as Alice and Bob keep either the long 'key' DNA or the chips safe, their messages will be safe from Eve's prying eyes. There are many other systems for DNA-based exchange of secrets; some are based on public-key

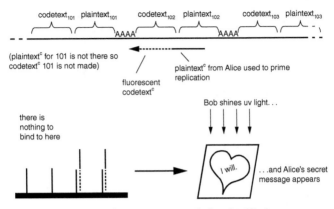

26. DNA encryption of images, part 3: Bob decodes Alice's message.

schemes similar to those used to transfer card details for online payments, for which Bob and Alice have no need to hold identical copies of a secret key.

DNA may change the face of cryptography in another way, based on DNA-based computing. Standard cryptographical techniques use a key to transform plaintext into ciphertext. In well-designed systems, calculating the key requires completely impractical numbers of messages to be analysed, leaving a brute-force method of trying all possible keys. Specialist computers have been built to crack systems such as the USA's Data Encryption Standard (DES) this way, but they relied on this standard using very short keys (56 bits): the use of much longer keys increases the computational load massively. As we have seen, though, tubes of DNA can offer trillions of computing elements that operate in parallel. Designs have now been published for DNA-based computers that would be expected to break 56-bit DES in days, and to cope with longer keys in not very much longer times. Presumably versions could also be constructed for other encryption schemes. It is unclear whether these have actually been built (the sorts of government agencies that build novel and expensive code-breaking machines tend not to broadcast their achievements) and it is not clear whether the error rate in DNA computing would be too great for them to work. DNA has a rival for solving large NP problems: quantum computing. If either could be made to work, they would greatly shift the balance of power between those who have secrets to keep and those who pry.

Chapter 6
Synthetic biology for basic research

Synthetic biology has been built on the foundations of a vast amount of fundamental knowledge about how genes, proteins, and living cells and tissues work. Having fed on this basic research, it has matured enough to begin to pay back its debt to science by providing sophisticated tools for analysis of natural systems, and for testing hypotheses and ideas. This chapter presents a few examples drawn from many.

Tools for enquiry

Biologists have long been frustrated by a mismatch of scale between the tiny size of cellular machinery and the comparatively huge scale of the instruments used to measure it. The electrical systems of neurons, for example, are based on nano-scale membrane proteins, yet measuring or initiating neuronal firing traditionally relies on electric currents in needles. The position of a needle is hard to control, its presence in a conscious animal may well interfere with normal behaviour, and there are obvious ethical concerns about stabbing needles into a higher animal's brain. One thing that experimenters can aim and analyse at micro-scales is light, but, unfortunately, most neural cells neither react to light nor produce it. Synthetic biology offers a means to place nano-scale protein-based devices in cells to translate light into neural firing or report neural firing as light.

To confer light-activated firing on neurons, Daniel Hochbaum and colleagues used the light-sensitive sdChR ion channel protein, from the alga *Scherffelia dubia*, as a starting point and engineered it to respond very quickly to blue light. When the engineered protein, 'CheRiff', is expressed in neurons it does nothing in the dark; however, when illuminated by blue light it opens an ion channel in the cell membrane, allowing ions to flow through and change the membrane voltage enough to initiate neuronal firing (Figure 27a).

Some natural light sensor proteins, such as bacterial archaerhodopsin, are weakly fluorescent and the strength of their fluorescence depends on the voltage across the membrane in which they are located. Natural archaerhodopsin has the awkward feature that illuminating it to measure its fluorescence causes it to alter membrane voltage. Hochbaum and colleagues therefore engineered it to report voltage without altering it and to react to changing voltage in less than a millisecond, making it as fast as traditional needle-based measurement methods. They called this protein QuasAr (Figure 27b). The experimenters

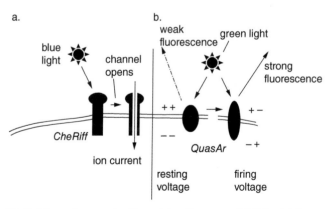

27. Driving and measuring the activity of neurons with light: (a) shows the action of the CheRiff light-to-voltage protein; (b) shows the action of the QuasAr voltage-to-light protein.

then bundled genes encoding both their light-to-voltage CheRiff protein and their voltage-to-light QuasAr protein as a synthetic biological module that could be introduced into animal cells. Placing the synthetic module in neurons in simple cell culture, and controlling and measuring them with either needles or light, showed that the synthetic biological device performed about as well as traditional ones, and of course had the advantage that they it required no needles. The system was then used to make real measurements of working neural circuits in a slice of brain kept in culture.

An ability to control the firing of specific neurons in living animals has been applied to test hypotheses about animal physiology and behaviour. *Caenorhabditis elegans* is a small, transparent roundworm with a nervous system so simple that its entire wiring diagram has been worked out. In 2011, two different laboratories published research papers describing synthetic biological devices built as tools for testing ideas about how the worm's nervous system functions. In both cases, the device conferred light-sensitivity on particular neurons, with both activation and inhibition of firing being possible via different colours of illumination. The researchers constructed computer-controlled microscope systems that allowed specific neurons to be tracked and targeted even in freely moving worms. The result of these collaborations between cutting edge machine vision specialists and synthetic biologists was effectively a 'remote-control worm': pulses of light could be used to fire neurons and steer the worm in any direction an experimenter pleased, for example travelling round and round a triangular path (Figure 28). The demonstrated connection between the firing of these neurons and the movement of the worm verified the provisional ideas about how the worm's nervous system worked. The idea has since been applied to other small animals such as fruitflies.

Optical control of neural firing can also be used to test ideas about the physiology of higher animals. One problem of significant

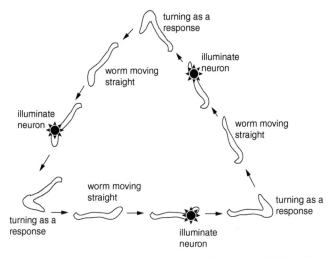

28. 'Remote control' of a roundworm using light-triggered firing of specific neurons.

societal importance is the physiological basis of addiction. Given that addictive behaviours are similar across a very diverse range of legal and illegal substances and activities, it was theorized many years ago that the drug addiction is not really to drugs themselves but rather to the effect that they have on the brain; a kind of internal addiction to the high levels of natural, pleasure-associated neurotransmitters made by the brain in response to the external trigger. One candidate for the molecule of 'internal addiction' is dopamine in the ventral tegmental area of the brain. In a recent paper, Vincent Pascoli and colleagues tested this idea by engineering into mice a synthetic biological module containing a protein that translates light into neural firing, coupled to a gene rearrangement system that ensured that it would be active in the dopamine-making neurons in the ventral tegmental area but nowhere else. The researchers also introduced, surgically, an optical fibre that could conduct light from an electronic device into that area of the brain. The mice were put into a living area in which they could press a lever, and the lever would communicate to the device and send pulses of light into their brains to activate the dopamine-producing neurons. The mice quickly learned to press the relevant active lever, ignoring other levers also present, and were soon reaching their daily eighty-press limit within the first hour. If pressing the lever was then associated with a mild electric shock, some animals avoided the lever, but others kept pressing it, something the experimenters saw as parallel to human addictive behaviour that continues despite negative consequences. One conclusion (there was more to this study than presented here) is that animals can indeed become addicted to direct stimulation of dopamine-producing neurons, suggesting that these may indeed be the mechanism of human addiction to substances such as cocaine that stimulate them.

Synthetic biology

The use of synthetic biology techniques goes far beyond neuroscience. Embryologists are often interested in the question of which cells in the early embryo make what structures in the late embryo or adult: answering this is important both to basic biological understanding and also to the development of techniques for tissue engineering and regeneration, because if one wants to rebuild a structure it helps to know where it comes from in the first place. Discovering what cells become is done by lineage tracing—introducing a permanent, undilutable (e.g. genetic) mark in one cell at one stage of embryonic development, letting development proceed, and then mapping which structures in the adult carry the mark. The process is laborious because it is essential that no more than one starting cell be marked, so tracing the lineages of many cells requires very many embryos.

A synthetic biological tool recently developed by Stephanie Schmidt and colleagues makes lineage tracing very much easier because it can work with many lineages in parallel. It makes use of the CRISPR system of DNA editing described earlier. The system, again tested using the roundworm *Caenorhabditis elegans*, involves introducing an inert 'victim' gene into the genome.

The CRISPR DNA-editing enzyme, and guide RNAs to target it to ten different sites in the victim gene, are injected into eggs of the worm. When it is active, the enzyme will cut the victim gene at one of these sites, and the cut will be repaired. When the repair is perfect, the original sequence reappears and can be cut again. Where the repair is imperfect, which it often is when CRISPR is used in this mode, an altered DNA sequence (a 'scar') will be left behind and, because this sequence no longer matches any of the guide RNAs, the site will now be left alone as a permanent, uneditable mark. Because the errors made in repair are essentially random, different cells will acquire a different set of scars (different in terms of which of the ten possible CRISPR targets are scarred, and different in terms of the exact error made at each scar). This gives each cell a unique 'barcode' that can be passed along to all of its daughters. As long as active CRISPR complexes remain, the barcoding continues so that daughters of one cell can add their own labels to pass on to their own daughters (Figure 29). If one dissects the tissues of the worm that develops

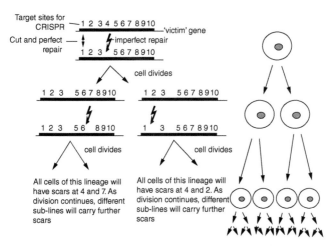

29. Stephanie Schmidt's barcode system: error-prone editing implants heritable DNA 'barcodes' in the cells of embryos, allowing the 'family tree' of cell divisions to be reconstructed.

from this embryo, and sequences the 'barcode' of each cell, it is possible to work back up its 'family tree' all the way to the egg.

It is striking that synthetic biological devices intended as research tools are now often published not in specialist synthetic biology research journals but in the mainstream journals of physiology, neuroscience, etc. A similar move happened many years ago to molecular biology, and it is a sign of a field's maturation from being an exciting curiosity to becoming an indispensable tool for other things.

Building to test understanding

In his 1912 book *La Biologie Synthétique*, Stéphane Leduc wrote 'when a phenomenon has been observed in a living organism, and one believes that one understands it...one should be able to reproduce this phenomenon on its own'. Decades later, the physicist Richard Feynman said 'What I cannot build, I do not understand.' Both quotations catch the spirit of an important use of synthetic biology: verifying that we understand biology as well as we think we do.

Discovery of biological mechanisms conventionally proceeds in three stages: description, correlation, and perturbation (sometimes in a different order). Description is just that—a careful study to establish the sequence of events when some biological process is going on. An embryologist might, for example, describe the sequence of anatomical changes that happens during the formation of a particular organ, and molecular biologists might catalogue what genes and proteins come and go in different cell types during the process. Correlations between these molecules and anatomical changes then suggest a hypothesis, for example that a mechanism involving proteins P, Q, R...drives event E. This hypothesis can be tested, to a limited extent, by perturbation: if one inhibits the action of protein P, or Q, or R, does event E fail?

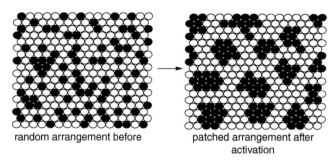

random arrangement before

patched arrangement after activation

30. A synthetic biological patterning system: on activation, it causes cells to make different adhesion molecules (symbolized by black and white cells in the diagram) and the adhesive forces organize the cells into animal coat-like patterns of patches.

This kind of hypothesis testing is adequate for proving that proteins are necessary for the event; however, it does not prove that they work in the way proposed and the complexity of the original system, where many other molecules will be changing, makes testing the mechanism difficult. Synthetic biology allows us to test the proposed mechanism by building a system in which *only* the proteins of the proposed mechanism, produced from synthetic biological constructs, change while everything else about the cells remains the same. In an early application of synthetic biology to embryological problems, Elise Cachat and I did precisely this to test a proposed mechanism by which mixed cells can spontaneously organize themselves into large-scale patterns (Figure 30).

A long-standing theory for the patterning of tissues in the embryo is Albert Dalcq's 1938 proposal that some parts of embryos secrete a signalling molecule that spreads out from them, forming a concentration gradient, with the fate of other cells being determined by the local concentration they detect (Figure 31a). Various embryologists have sought this mechanism in real embryos and have found several instances that seem to match. The complexity of the embryo does, however, make it

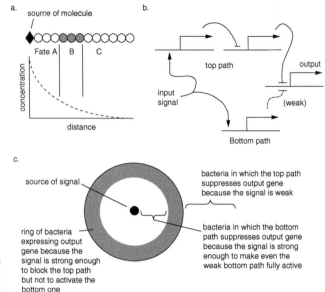

31. **A gradient of signal (a) can be interpreted by the synthetic biological network shown in (b) to control bacterial gene expression according to distance from a source of signal (c).**

difficult to be certain that this simple system is all that is needed to account for the pattern. To show that the idea can work in principle in real biological systems, Subhayu Basu and colleagues, and (separately) David Greber and Martin Fussennegger, engineered bacteria with synthetic three-gene networks inspired by gene network architectures found in embryos. In each case, the network detected a small molecule rather than a real mammalian signalling molecule, and activated a reporter protein rather than a developmental event: this meant that the basic patterning function of the network could be tested in the absence of other complications. To take the Basu construction as an example, the network involved two paths from the signalling molecule to the output gene (Figure 31b). The top path was a 'double-negative': the signalling molecule caused the expression of a gene that inhibited the expression of the next gene, that would in turn inhibit the expression of the final output. In the absence of the signal, this top path would inhibit the output gene; while in the presence of the signal it would not and the output would be on. In the bottom path, the presence of the signal would switch drive the expression of a gene would inhibit the expression of the output gene. Crucially, the bottom path involved a rather inefficient inhibitor that needed a high concentration of the signal to work well. There was therefore a range of signal concentration in which there was enough signal to drive the top path sufficiently to allow expression of the output gene but not enough to drive the bottom path sufficiently to inhibit it. Thus populations of bacteria harbouring this construct, grown on an artificially applied concentration gradient of the signalling molecule, produced a stripe of output gene expression where the concentration was moderate but not where it was high or low (Figure 31c).

Using synthetic biology in this way adds an extra stage to investigation, making the whole sequence 'description, correlation, perturbation, and synthesis'. In principle, the approach can be applied to a vast range of problems. It should be noted, though, that all that the synthesis step shows is that an idea drawn from a natural system does work *in principle*. It does not actually prove that the mechanism is (all that is) driving the natural system and wise minds will remain open to the possibility of unknown unknowns. In biology, there is almost always room for reasonable doubt.

A cornerstone of biology is that the character of a species is determined by both its genome and its environment; therefore, if the environment is the same, differences between two species should be determined by the genes alone. This hypothesis was tested in a bold experiment of 2010 by Dan Gibson (he of 'Gibson assembly', Chapter 2) and his colleagues. They assembled a complete copy of the genome of the tiny bacterium *Mycobacterium mycoides*, taking the necessary information from an electronic record of the bacterium's DNA sequence rather than copying physically from its actual DNA. This was a tour-de-force of synthesis because, although the bacterium has one of the smallest genomes known, it still required the assembly of over one million bases and very extensive quality control steps during the process. The synthetic version included a few minor edits, including the addition of 'watermarks' to aid identification. When the genome was ready, it was transplanted in to the cell of a different species—*Mycoplasma capricolum*. The recipient bacterium grew well, had the watermarks, and behaved like *Mycobacterium mycoides*, exactly as theory predicts. It was nicknamed *Synthia*, and it achieved much media coverage at the time because it was widely misunderstood by journalists as the creation of life (it was emphatically not that—the synthetic genome was after all put into a living cell). This misunderstanding was not helped by some unfortunate statements from some of Gibson's senior colleagues.

An example of a much larger scale synthetic biology experiment designed to test hypotheses about genomes is the Yeast 2.0 project. The aim of this international collaboration is to replace all of the natural chromosomes of yeast with synthetic versions. The replacements are generally edited versions of the natural, with pieces of 'junk DNA' (DNA with no discernible function) removed. Testing the viability of the resulting cells provides a strong test of whether these regions of DNA really are junk or whether they are in fact needed for some aspect of biology we so far know nothing about (which would be the more interesting result). The edited replacement genome includes some synthetic biological devices to enable rapid change and evolution of the yeast, allowing theories about evolution and adaptation to be tested. There is also a plan to include an artificial yeast chromosome, being built in Edinburgh, to host all of the genes encoding yeast tRNAs, allowing these genes to be removed from yeast natural chromosomes. This is partly to test some hypotheses about the positions of the natural tRNA genes (essentially, whether their natural positions, often near unstable

DNA, is important). It will also allow the engineering of tRNAs to make yeast strains that can incorporate new amino acids in addition to the twenty naturally used.

For basic science, the failure of a design abstracted from studies of real life to work properly when realized as a synthetic biological system will not be a 'failure' in the way it would be for engineering. Science often makes its greatest advances when things that 'should' work do not (measuring Earth's speed through the ether by interferometry 'should' have worked according to 19th-century physics; trying to explain its failure led to Einstein's theory of relativity). 'Failures' warn us that our analyses were wrong, and that there is still something new to discover. As biochemist-turned-science-fiction-writer Isaac Asimov observed, the most exciting thing to hear in a laboratory is not a shout of 'Eureka!' but a quiet mutter of 'That's odd!'

Chapter 7
Creating life

Earlier in this book, I introduced the idea of synthetic biology having 'two souls', one focused on engineering new features into existing life and the other aiming to create life *de novo*. Though the first attracts the bulk of current effort and funding because of its industrial, environmental, and medical potential, the goal of creating life is arguably far more significant in terms of long-term scientific and philosophical impact. It is poised to be another step in the long journey from Copernicus (the Earth is just one part of the universe, not its centre) and Darwin (humans are animals, and share their origins) to an irrefutable demonstration that life is just normal matter, albeit organized in a particular way. Most scientists already take this as a working assumption; however, it has not yet been proven, and it will not be until someone creates life from non-living matter.

Reasons for creating life

The project of creating life *de novo* is important for at least three scientific reasons. One is refutation of the vitalist hypothesis. A second is that creating life offers a way to test hypotheses about the origin of natural life by simulating the conditions of early Earth in the lab. The third reason is that trying to create life may teach us much more about what life actually is: our current answers to that deepest of biological questions are still largely untested conjecture.

There is a broad (but not universal) agreement that, to be classified as living, an entity must show the following four features:
(i) metabolism, to make use of both raw materials and energy;
(ii) stored information, to specify structures and functions;
(iii) containment, to keep everything together; and
(iv) reproduction, at least at the level of a population.

These criteria are deliberately phrased in an abstract way that does not require any specific type of structure or chemistry, and they could be applied equally to a candidate for alien life.

There are two very different approaches to making life *de novo*. The 'soft option' is to take a minimal combination of proteins, genes, etc., from existing organisms and to combine them in some kind of membrane so that, if the parts have been chosen wisely, the system will be self-sustaining and will reproduce. This approach is useful to test whether we have an adequate understanding of cell physiology, but it is not useful for deeper questions because it borrows too much from existing organisms. The more extreme approach is to create life from the bottom up, starting with simple chemicals and taking nothing from existing living things except inspiration. In the interests of space, this chapter will concentrate only on this more ambitious goal.

Inspiration from assumed origins

The model that has dominated origin-of-life research for almost a hundred years was first stated clearly by Alexander Oparin, in 1924. It begins with chemical reactions between very simple gases and salts to produce the first organic (i.e. complex, carbon-containing) molecules. The feasibility of this step was shown by Harold Urey and Stanley Miller in the 1950s, and those who have built on their work. They exposed inorganic gas mixes representing the atmosphere of early Earth, and water representing its oceans, to warmth and 'lightning' sparks. The result was the production of a rich variety of organic molecules including amino acids, sugars, and nucleic acid bases. The Oparin story then features a gradual increase in chemical complexity, as small molecules reacted with one another to make larger and more diverse structures. The reactions to make these may have been catalysed by minerals and clays, which have been shown to have catalytic activity. Some of the larger molecules would themselves have bound smaller ones well enough to show at least a weak ability to catalyse reactions between them. So far, this was 'pure' chemistry, with no higher organization.

In Oparin's theory, the first organization emerged when a set of catalysed reactions happened to coincide, in which the synthesis of each chemical component from its precursors was catalysed by another molecule in the set (see Figure 32 for a stylized example). In principle, as long as some method of containment existed to keep the set together, and as long as starting materials could enter, such a system could make more of itself: that is, it could 'reproduce'. It is important to note that the molecules involved would not necessarily be the ones common in living cells now: the Earth's environment has changed a great deal since the origin of life, and four billion years of evolution and adaptation will

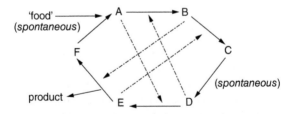

32. **An illustration of a catalytic cycle, in which each slow (non-spontaneous) reaction is catalysed (dotted arrows) by a specific product of the cyclical pathway.**

probably have seen almost all of the original components replaced with something more appropriate as conditions changed. The fact that the parts of Concorde are not interchangeable with those of the Wright Brothers' first aircraft does not mean that one does not owe its origins to the other.

Containment is clearly a critical factor in a system such as that in Figure 32, to keep the system together. Closed membranes will form spontaneously when molecules with hydrophilic (water-liking) heads and hydrophobic (water-hating) tails are suspended in water to form structures such as micelles and vesicles (Figure 33). What is more, there is a natural maximum size to these structures and, if they grow too large, they spontaneously split into two small ones, in much the way that long soap films blown from hoops break up into individual soap bubbles. In principle, then, one could imagine having a catalytic cycle as described in the last paragraph enclosed in a membrane vesicle; as long as the catalytic cycle included synthesis of more

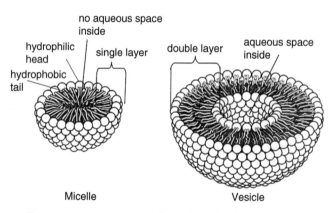

no aqueous space inside

hydrophilic head

single layer

double layer

aqueous space inside

hydrophobic tail

Micelle

Vesicle

33. Two common arrangements taken up by molecules that have a hydrophilic head and a hydrophobic tail, when they are suspended in water. In both cases, the hydrophilic head is in contact with water and the tail is not. Micelles are always small and have no watery interior; vesicles may be much larger than the one illustrated, and include a watery interior which might also include a metabolism or nucleic acids.

membrane molecules, and as long as the membrane admitted essential reactants, one would have a self-sustaining and even reproducing system. Of the criteria for life mentioned earlier, it would clearly possess metabolism, containment, and reproduction. How about information? Such a system does not have genes; however, each of its molecules has a structure and, in biology, structure is information (even for genes). The structures of the reactants coupled with the shapes of the catalytic components specify the nature of their reaction products; in principle a different shape, caused by a 'fluke' reaction, could specify a different product and a different version of the proto-organism, opening the door to evolution. So there would be information in these systems, but, unlike genetic information, it would not be held separately from the metabolic activities of the cell.

In this 'metabolism-first' story of life's origins, how might genes have arisen? One popular story is that RNA first arose in the catalytic cycles of these evolving early organisms as a catalytic rather than a genetic molecule; RNA molecules have been shown to be able to act as catalysts. Later, RNAs that bound amino acids at one end and another RNA at the other were used to bring these amino acids together leading to the first examples of peptide chains being made using amino acids specified by RNAs. The great catalytic efficiency of peptide chains and the proteins they form meant that protein-based catalysis gradually took over from earlier catalysts. Eventually, proteins copied RNA, which made proteins. Thus RNA-based genes, now set apart from metabolism and playing only an information function, entered the system. The use of RNA as a genetic material is entirely feasible: many modern viruses do it. Separation of genetics from metabolism allowed DNA, a more stable molecule for long-term information storage but one that is catalytically fairly useless, to replace RNA as the main genetic molecule. Thus the story ends with life as we know it.

The highly abbreviated origin-of-life story just summarized is
not the only one: in particular, there are versions in which RNA
comes first and metabolism and containment comes later. The
version described does, though, provide a convenient framework
for description of synthetic biological attempts to create life in
the laboratory, which generally set out to follow the same path.

Synthetic catalytic cycles

The search for cyclical reaction schemes that can transform simple
precursors into more complex organic molecules is an old one.
The first clear example was the formose cycle, discovered by
Aleksandr Butlerov in 1861. The reaction (Figure 34) begins with
glycoaldehyde. A molecule of formaldehyde joins this to make
glyceraldehyde, which isomerizes to form dihydroxyacetone. Yet
another molecule of formaldehyde joins this to make ketotetrose,
which isomerizes to make aldotetrose. This splits to produce two
molecules of glycoaldehyde, each of which can begin a new cycle

**34. The formose reaction, in which a cycle turns formaldehyde into
glycoaldehyde via more complex molecules, and seeds new cycles. The
cycle is therefore, in a sense, self-replicating.**

of its own. Thus the net effect is to transform formaldehyde into glycoaldehyde, via larger sugar-like intermediates, in a cycle that reproduces itself in an autocatalytic cycle.

This simple cycle is a long way from a full cellular metabolism, but it, and a few more recently developed ones, shows that self-reproducing cycles are at least possible.

Making compartments

Among the most interesting catalytic systems to have been explored in the context of synthetic life are those that produce membraneous vesicles as their end-product, because such vesicles are excellent candidates for containment of metabolic systems. In the early 1990s, Bachmann and colleagues floated the oil ethyl caprylate over alkaline water. The alkali slowly hydrolysed the oil to make sodium caprylate, which formed micelles in the water. The surface of the micelles acted as a catalyst for further hydrolysis of the oil, creating an autocatalytic system in which micelles begat further micelles (Figure 35).

35. The self-catalysed reproduction of micelles of sodium caprylate: the starting compound is ethyl caprylate, which will floats above water. It is slowly converted by sodium hydroxide, dissolved in the water, into sodium caprylate, which makes micelles. The micelles accelerate the process, so that micelles cause the formation of more micelles.

Micelles are not very useful as containers because they have little internal space (Figure 33): more useful are vesicles formed of membrane bilayers. Under the right conditions, micelles can make vesicles. Sodium oleate, like the sodium caprylate, has a hydrophilic head and a hydrophobic tail, and forms micelles in aqueous solution. At pH 8.8, these micelles slowly fuse and rearrange to become a vesicle containing an aqueous interior. The presence of a vesicle catalyses the conversion of further micelles into more vesicle membrane, so that vesicles grow. When they reach a critical size, they split to form two small vesicles and the cycle continues (Figure 36). This mechanism does not involve production of complex molecules from simple ones (it begins with the oleate), but it does involve a cycle in which large multi-molecular structures form and reproduce from more simply arranged precursors. Solutions of myristoleic acid also form vesicles, slowly on their own but much more quickly if particles of clay are added: in this case, the vesicles encapsulate the clay. If nucleic acids are first bound to the clay, they become encapsulated too. This raises an interesting possibility for a method by which partially clay-catalysed or clay-bound catalytic cycles might become encapsulated into a compartment. What does not yet seem to have been done is the encapsulation of a

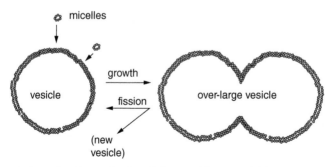

36. Growth and reproduction of oleate vesicles; growth is by recruitment of new oleate from micelles, and reproduction by fission of large vesicles.

metabolic cycle that replicates all of its components into one of these self-replicating compartments. That would be a major advance.

First steps to genes

At least in the metabolism-first versions of the story of life on Earth, genes came late and were not a feature of elementary life. Many synthetic biologists ignore them as belonging to a biology far too complicated to be the concern of people wanting to create minimal life. Others, though, question how complicated a very basic genetic system has to be, and have asked whether it is possible for an information-containing molecule to replicate itself in a template-directed manner without the paraphernalia of proteins. One way in which a nucleic acid strand might act as a template for its own replication would be to act as a docking site for a complementary base or small sequence of joined bases, which will be held side-by-side by the interaction. Under reaction conditions that allow low but non-zero probability of the joining of these subunits, joining in bulk solution would be infrequent because chance meetings between the molecules will be infrequent; however, when they are held side-by-side for long periods, joining will be more likely and the complementary strand will be assembled (Figure 37). Subsequent dissociation of the two strands (e.g. by heating in a day–night cycle) will make each available to be used as a template for its own replication.

This type of non-enzymatic replication has been demonstrated for a few specific types of nucleic acid. In all cases the replication works only for short lengths. In one example, a six-base peptide nucleic acid (i.e. a nucleic acid with a backbone made of peptide rather than the 'normal biology' phosphodiester bonds) acted as a docking site for two three-base nucleic acids, and held them together for long enough that they joined. A similar system has been built with normal, phosphodiester-linked nucleic acid. Further experiments have been conducted using more than one template, and different sequences have shown competition;

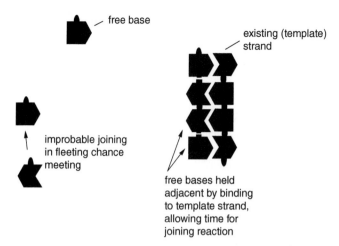

free base

existing (template) strand

improbable joining in fleeting chance meeting

free bases held adjacent by binding to template strand, allowing time for joining reaction

37. **The idea of template-directed synthesis of nucleic acids without the use of enzymes.**

'evolution' in the system is therefore possible. Self-replicating protein chains have also been constructed, in which the chemical activity of the protein allows it to assemble more of itself from two shorter subunits.

It is important to note that the only aspect of a 'gene' captured by these experiments has been template-directed replication. Unlike real genes, the nucleic acids concerned do not direct the production of anything but themselves. The main point of the work is to tackle the chicken-and-egg problem of whether nucleic acids or proteins came first, since in normal biology each is only made with the help of the other. At least under rather special conditions, the experiments have shown that some nucleic acids can self-replicate without the help of proteins, and that some proteins can self-replicate without the help of nucleic acids (in both cases from some rather complex precursors, not from simple 'food'). One could therefore imagine an evolutionary sequence in which nucleic acids such as RNA existed at first parasitically in the metabolism of early cells, and those that happened to mutate and acquire useful enzymatic activity conferred extra fitness on the cells; parasitism became symbiosis and ultimately union into one coherent system.

Next steps

So far, laboratories working on creating life from scratch have generated self-reproducing catalytic cycles that generate molecules of modest complexity, such as sugars, from simple precursors such as formaldehyde. Those working on containment have made self-reproducing membrane vesicles, albeit ones that require somewhat complex starting molecules. What seems not yet to have been done is to join these concepts so that a self-reproducing catalytic cycle can begin with very simple precursors, make molecules that can make membranes, and so 'feed' the self-reproduction of the vesicles. If this can be achieved, then the next step would be to encapsulate the cycle in the membrane, and to ensure that the small starting molecules could get in. Such a system would be a very great step towards making a minimal living thing. Many people would consider such a thing to be alive. Others, fixated on the idea that life has to include genes, would not.

It is possible that any philosophical ramifications of humans having created life will be cushioned by a long period in which definitions of what it is to be alive change. Perhaps the creation of synthetic life will be recognized only in retrospect, not as the outcome of one critical experiment but as a series of small steps—steps that crossed what turned out not to be a sharp boundary but rather a fuzzy zone of ambiguity between the inanimate and the living. Such a zone of ambiguity does, after all, exist at the boundary between being alive and being dead: in a multi-cellular organism such as a human being, individual cells live on long after one would consider a person as a whole to have died.

Chapter 8
Cultural impact

Science is not an isolated enterprise: it is affected by, and in turn affects, broader society and culture. Synthetic biology has changed aspects of education; has stimulated artists, writers, and film-makers; and has caught the interest of philosophers, ethicists, campaigners, and legislators. This final chapter gives a taste of the wider implications of the technologies described earlier in this book.

Education

The rise of synthetic biology is having an interesting effect on biology education, catalysed by the International Genetically Engineered Machines (iGEM) competition. Founded by the Massachusetts Institute of Technology in 2004, iGEM features student teams who are sent a kit of synthetic biological parts and who design and build synthetic biological devices that use them. When teams design new parts they add them to the kit for next time. Hundreds of student teams across the world enter and there is a grand final in Boston each autumn. The experience of working on a project of their own—designing, building, and testing their device, with accompanying ethical and safety assessments and preparation of detailed public reports—gives students a range of experiences far beyond what they would get in a lecture theatre. Most projects are of very high quality: the first version of the arsenic detector featured earlier was, for example, designed by an iGEM team. As well as learning the techniques of manipulating biological systems, the students learn the value of imagination (something inadequately stressed in conventional science teaching) and of working with team members from other disciplines.

The culture of iGEM has inspired similar initiatives in other fields. One of the most culturally interesting is the Biodesign Challenge, intended primarily for students of art and design. The scheme has themes including architecture, communication, energy, food, water, materials, and medicine. Teams have to make well-designed proposals, and proper models of the bio-designs they propose; however, unlike iGEM teams, they do not have to build the devices. The entrants can therefore think at much larger scales of space and time than iGEM teams (who have to build something in a matter of months), but feasibility is a key criterion in judging. Winners' designs are exhibited in an annual Biodesign Summit in New York. Initiatives such as this show great promise in producing educated young people who see no great barrier between C.P. Snow's 'Two Cultures' of arts and sciences.

Art

Artists are often among the first people to respond to a new technology. Artworks using synthetic biology as a medium are varied, most using unicellular life for practical and ethical reasons. One example from C-lab (an artists' collective founded by Howard Boland and Laura Cinti) consists of plates of light-responsive bacteria that record the image of a human face in a living medium rather than in the inanimate chemicals of photographic film. The image is not only recorded by life: it is also altered, blurred, and finally erased as individual bacteria move, multiply, or die, much as the real face will one day be lost to unavoidable processes of biological change. In another installation, bacteria in growing colonies report their own stress as an intensity of fluorescence; the coexistence of stressed members of the colony and unstressed 'freeloaders' is made visually obvious, provoking thoughts of the city in which the exhibit is contained. Yet another example from the same collective uses engineered bacteria to 'paint' a landscape of smells, allowing the artist to engage effectively with olfaction which is, in art terms, a largely neglected sense.

Some 'synbio-art' installations are designed for interaction. An example is the 'Genesis' project of Eduardo Kac, who encoded into DNA the quotation from Genesis in which God, blessing Noah and his sons, gave them dominion over every beast of the Earth. Kac engineered this text-bearing DNA into living bacteria. Viewers of the exhibit were asked to decide whether they agreed that humans should have dominion over nature; if they did not, they could activate an ultraviolet light that would subject the DNA sequence to random mutation, destroying the integrity of the quote. But to do so, they had to exercise their own power over the life in the exhibit.

Some works have been produced as a result of programmes in which synthetic biology laboratories host artists-in-residence: this habit is another example of synthetic biology being 'different' among the sciences.

Science-fiction literature and film also show rapid responses to the new. By the late 1980s, science fiction was acquiring a distinct new sub-genre, 'biopunk'. Biopunk novels tend to be set in dystopian worlds peopled by biohackers, unethical mega-corporations, oppressive police forces, and black-markets in illegal body modification. The lead characters ('heroes' does not really fit) tend to be near the bottom strata of a very unequal society, with no hope of rising by legal means. In thirty-five bio-fiction films recently analysed by Angela Meyer and colleagues, most synthetic biologists were portrayed as being motivated by entrepreneurism of various kinds, or were just a small cog in a big corporate machine. This change of cliché, from the academically driven, lone, mad genius of the black-and-white era of science fiction to the incompetent or unethical tool of a capitalist corporation, may reflect how society's view of a typical scientist has changed over the decades. Few of us would consider it flattering.

Recreation

The rise of 21st-century synthetic biology has nurtured the development of a new, molecularly based biology hobby, 'DIY-bio' or 'biohacking', to join the far older biological hobbies of natural history, gardening, and pet breeding. As 'biohacking' implies, the hobby consciously draws on the spirit of the DIY computing community in which hobbyists construct and experiment with small computers using easily available parts and knowledge that is freely exchanged. Although done for fun, the results of this tinkering can be astonishing: the micro-computer industry that is now such a major part of world economics and culture was spawned largely by amateurs in sheds and small clubs. In both communities, 'hacker' means someone with a gift for improvised engineering and repurposing, and has no connotation of nefariousness (that meaning arises from a press misunderstanding).

DIY-bio has grown mainly as a network of clubs that build their own laboratories, sometimes as a part of a larger 'maker-space' that serves other practical hobbies. Equipped with second-hand apparatus, often donated by universities and industry, or with cleverly improvised equipment, these labs are modest; however, their builders have confidence that, as in the world of computing, the imagination of enthusiasts will result in interesting inventions. DIY-bio has to operate within a legal framework that regulates genetic manipulation, environmental release of manipulated organisms, and animal experimentation. Other technical hobbies also operate within strong regulatory frameworks (amateur radio and amateur rocketry are examples) and, in some cases, hobbyists require a licence to operate, obtained following a formal assessment of competence. Currently, none of the countries in which DIY-bio is common have introduced compulsory licensing, but the community itself has created rich educational resources on safety and responsibility. The DIY-bio communities in Europe and the USA have clear 'codes of ethics', which include transparency, safety, open access, education, peaceful purposes, responsibility, and accountability.

Given the restrictions on what work can be done, and perhaps given the fertile imaginations of the people involved, DIY-bio seems to be following a different path from mainstream synthetic biology. There is much more emphasis on art, and much construction using plants and microorganisms is done by influencing the cells at the epigenetic level (i.e. without altering their DNA sequence at all and therefore without becoming entangled in regulations). In some places, there is an emphasis on creating 'cyborg' constructions; for example, robots whose behaviour is controlled by the behaviour of an insect in a living space within them. A small but attention-grabbing faction of the biohacking community takes the cyborg idea further: calling themselves 'grinders', they aim to biohack themselves towards a state many call 'transhuman'. At present, synthetic biology is too young a science, and frankly too difficult, to have made much of an impact on the movement and most 'grinder' activity involves people implanting electronic devices into their bodies. The movement communicates ideas in forums (e.g. biohack.me) and design repositories (e.g. grindhousewetware.com). At least one biohacker has undergone limited genetic modification, introducing DNA intended to produce growth-hormone-releasing-hormone into his own body, in an attempt to extend his lifespan. It is not clear how many more people have tried this type of thing, or how many more will in the future as the required technologies become easier to obtain. The ethico-legal background to this type of self-experimentation is complex and unclear, as legal developments are outrun by technical ones.

Ethics

New developments in biotechnology tend to spark ethical debates: birth control, heart transplantation, diagnosis of brain death, and even the use of anaesthesia have all been condemned as unethical by some. Synthetic biologists have deliberately invited measured ethical discussions having learned, from the fiasco around the public reaction to GM crops in the European Union, what happens when debate is delayed until advances have already been made and is then dominated by chaotic and ill-informed public panic.

Ethical discussions are complicated because different people take different approaches. Consequentialists judge the ethics of an action according to its effect (for them, even a well-intentioned act that ends up doing harm is morally bad). For most consequentialist thinkers, there is no judgement to be made about synthetic biology as a way of working, and each individual project will be judged on its likely and worst-case outcomes. Deciding on the balance of benefit and harm may be difficult in practice, as even something such as cheap and nutritious food that might seem at first sight a clear 'good' may be massively disruptive to the lives of traditional peasant farmers. Most consequentialist arguments remain largely in the realms of safety, well-being, and economics, and run along the same lines that currently control the development of genetically modified organisms, drugs, and chemicals. There is one strand of consequentialist thought that can lead to a conclusion that synthetic biology is bad in principle: it is the concern that viewing life as something to be manipulated makes us devalue it, and therefore treat lives, including our own, with less respect. This may apply both to engineering of existing life or to building life *de novo*. Previous major steps that dethroned Earth and Man from having a special status in the universe (the work of Copernicus and Darwin) were followed by significant social change (e.g. the Reformation and Fascism). This may be coincidence or arise from the selection bias of historians trying to find patterns in random noise, but at least some commentators feel that deleterious socio-political responses to man-made life are a real possibility. One major challenge for consequentialist thinkers, particularly those designing regulations, is to decide how far to look ahead when making a decision about what should be allowed now. Do we have to try to see everything that can follow, or just look forward to a next milestone that will trigger a fresh round of debate?

The other main strand of ethics, deontology, focuses on the morality of intent (the only thing that can be unqualifiedly 'good' is good will, as Immanuel Kant put it) and the morality of the thing that is done, rather than on its consequences. To a deontologist, a malign or morally unacceptable act that accidentally does a lot of good is still a bad act. Application of deontological thinking to questions of synthetic biology usually hinges on deep values, such as the inherent merit of knowledge and whether life and nature should be held 'sacred' (literally, or some secular equivalent). Here, creation of life may be the less contentious activity: deontologists are unlikely to find much moral dimension in any specific manipulation involved in engineering protocells, as these are just the routine tasks of a chemistry lab. Manipulation of existing living things to suit our purposes is more of an issue, and comes down to the question of whether there is a boundary between the natural and artificial that scientists should not cross: whether it is intrinsically wrong to manipulate life and use it merely as a means to an end. Where the living things concerned are animals, deontological considerations about synthetic biology will merge with those about animal rights. Deontological objections to synthetic biology are common in the environmental movement, although for public discourse they are often disguised behind consequentialist arguments because these are widely perceived as being more likely to sway public opinion.

Most major religions apply to synthetic biology the same ethical frameworks they have long applied to manipulation of the physical world and to existing medical interventions. Like many non-religious people, they tend to focus on consequentialist tests of whether what is done is good for human well-being, dignity, and respect or whether it works against these things. For many religions, including the three Abrahamic ones, humans are already partners in the work of creation and synthetic biology is not crossing any great ethical line.

Ethics are translated into practice as regulations, and laws. When written well, regulations work for everyone: society as a whole is protected from technologists doing what has been prohibited, and technologists working within the regulations are somewhat protected from accusations of dangerous irresponsibility. Approaches to legislation vary hugely. Some pressure groups, particularly those espousing 'green' politics, advocate the 'precautionary principle', which argues that new technologies should be banned until they are proven to be harmless. The precautionary principle appears to be obvious good sense, but, in reality, it is almost impossible to prove something is harmless until it has been built and tested; the precautionary principle is, therefore, effectively a ban on anything new. Most legislatures take a pragmatic approach and require formal assessment of risk and, for some types of work, formal permission is needed from a government or independent agency. In many cases, the rules are centred on techniques used so that, for example, environmental release of a plant made by conventional breeding would not require a licence but one made by genetic modification would, even if the resulting genome has exactly the same change in sequence. This approach repeatedly runs into the problem that new techniques do not fit neatly into the frameworks of existing laws. Canada is an example of a country with a quite different approach, in which licensing is based on the properties of the thing made rather than on the technique used to make it. This framework accommodates new methods without any problems. Many areas of the world have little or no legislation, but that does not mean that synthetic biologists go to those places to do what they like. The delay in application of the arsenic sensor described earlier has arisen because of a conscious choice not to use a product in the developing world until it has been licensed within the European Union, because doing so would in some way imply that citizens of the developing world were worthy of less protection than Europeans. The delay itself does, of course, carry negative consequences for people still exposed to arsenic.

Legislation controls not only what can be made but also who owns it. Ownership of the product itself, of the design, and of the methods used to make it are all issues of great importance to commercial exploitation in a capitalist economy, and there are great tensions between those who argue that information should be proprietary and those who argue for total openness. The decision is not only economic—it is again ethical, because of what it means for the way we look at life. As Leon Kass commented, 'It is one thing to own a mule; it is another to own *mule*.'

Fear

Regulations are effective only on those who choose to obey them. Much speculation has centred around the possibility of synthetic biology being 'weaponized' to create synthetic pathogens. That 'rogue states' might contemplate bio-warfare is not in doubt: Americans have already suffered biological attacks from that historically most roguish of nations, Great Britain, which used smallpox against Native Americans in 1763 and probably against the American Militia near Quebec in 1775. The British also developed and tested Anthrax bombs on Gruinard island in 1942 although, fortunately, these were never used on their intended German target cities. In more recent years, bio-weapons have been used by terrorists, with attacks having taken place with natural *Salmonella* (USA, 1984), and *Bacillus anthracis* (anthrax: Japan, 1993; and USA, 2001).

Concern about synthetic bio-weapons has prompted government departments, such as the FBI's Weapons of Mass Destruction Directorate, to arrange for some of their staff to undertake training in synthetic biology labs to gain foundational knowledge in the new field. The UK's Home Office and the USA's Department of Homeland Security have also hosted meetings between synthetic biologists and representatives from various three-letter agencies. Taking place in locations so varied that some were like a set for *Yes, Minister* and others were like a set for *Dr Strangelove*, the conversations that this author experienced all came to a reassuring conclusion (at least for synthetic biologists): if a terrorist group wishes to cause havoc and mayhem, there are much easier and more certain means than those offered by synthetic biology. Being a successful pathogen is very difficult and, of all of the countless millions of bacteria and viruses in the world, only a tiny fraction are dangerous. To create an epidemic, a pathogen's biology has to be balanced finely enough that it can multiply in one host to numbers capable of infecting on average at least one other person before it either kills or is killed by its host. Most microbes do not have this optimal virulence; they are either killed before they spread or, occasionally, they are so virulent that they make people too sick to socialize and pass on the disease. Human variation ensures that individuals have different abilities to fight any given pathogen and the presence of a large number of resistant people makes it difficult for pathogens to spread between susceptible people (this is the basis of 'herd immunity', and the reason that vaccinating *almost* everyone is usually good enough). We still have only a very incomplete understanding of how virulence is controlled, so the idea of building a custom virus or bacterium with optimum virulence is far-fetched. The production of new and dangerous strains of influenza, described earlier, used random mutations of an existing virus, not direct design. Synthetic biology can indeed provide tools to make this kind of random breeding more efficient, but it is so easy to obtain naturally evolved and dangerous pathogens from the parts of the world where they are endemic that a would-be bioterrorist would be more likely to do that instead.

There is occasional speculation that synthetic biology might be used to adapt a natural pathogen to be active only against a specific human race. This is based on a misunderstanding of our species; humans do not have separate 'races' in the sense that a geneticist would recognize. Our species is a continuum: some versions of genes (alleles) do turn up at different frequencies in different ethnic groups, but this is a statistical property only. There are no absolute genetic differences that separate peoples on opposite sides of even a 'racial' conflict that would make one side vulnerable and the other safe.

Bioterrorism aimed at a nation's agriculture may be a more realistic worry. One of the main defences we animals and higher plants have against bacterial or viral Armageddon is sex. We reproduce not by copying ourselves but by shuffling and mixing our genes with those of another individual so that our offspring differ from us and from one another. This variation means that epidemics do not kill everyone. In a world of clones, in which all individuals were the same, there would be no herd immunity and a pathogen that evolved or was engineered to infect one individual would have a high chance of spreading through the population like wild-fire. This is essentially what happened in the Irish potato blight. If we are stupid enough to grow clones of crops in which field after field contains plants with exactly the same genomes and/or the same synthetic devices aboard, then we will be making our civilization far less resilient against fungal, bacterial, or viral attack, whether natural or deliberate. The current, very foolish, use of the same computer architecture and operating systems in everything from consumer goods to safety-critical industrial systems has allowed computer malware writers to cause much trouble in recent years. Had we still the diversity of small computers seen in earlier decades, we might have had to bear more cost in software development but our systems would be far more resilient in the face of mischief. We need to avoid making the same mistake with biology.

Hope

It would be misleading to end this book on a note of fear. Most speculation about synthetic biology takes an optimistic view of the good that may come from an ability to engineer living things. In the near future, there is much that can be done for the fields of energy, environment, medicine, and engineering. There is also hope that a better ability to shape life will reduce our need to over-exploit rare and endangered biological resources of our planet. Looking much further ahead, there is speculation that synthetic biology might allow us to bring life to lifeless planets and perhaps to realize some of the science fiction dreams kindled by previous new technologies. Beyond these practical speculations and desires, there is a more spiritual hope: that through creating life we will understand more about ourselves. For many thinkers, better understanding of our own natures and our place in the universe is in itself a good far stronger than mere material gain.

Further reading

Modification of existing life

Armstrong, R. *Living Architecture: How Synthetic Biology Can Remake Our Cities and Reshape Our Lives*. TED books/Amazon (Kindle edition only at present).

Freemont, P.S. and Kintney, R.I. *Synthetic Biology—A Primer* (Revised Edition). London: Imperial College Press, 2015.

Kuldell, N. *Biobuilder: Synthetic Biology in the Lab*. Boston, MA: MIT Press, 2015.

Regis, E. and Church, G.M. *Regenesis: How Synthetic Biology Will Reinvent Nature and Ourselves*. New York: Basic Books, 2014.

Schmidt, M. *Synthetic Biology: Industrial and Environmental Applications*. London: Wiley, 2012.

Creation of life *de novo*

Luisi, P.L.L. *The Emergence of Life*. Cambridge: Cambridge University Press, 2006.

Rasmussedn, S., Bedau, M., Chen, L., et al. *Protocells: Bridging Living and Non-Living Matter*. Cambridge, MA: MIT Press, 2009.

Culture, ethics, and art

Ginsberg, A.D., Calvert, J., Schyfter, P., Elfick, A., and Endy, D. *Synthetic Aesthetics: Investigating Synthetic Biology's Designs on Nature*. Cambridge, MA: MIT Press, 2017.

Jorgensen, E. *Biohacking—You Can Do It, Too*. TED talk. Available at: <https://www.ted.com/talks/ellen_jorgensen_biohacking_you_can_do_it_too>, nd.

Kaebnick, G.E. and Murray, T.H. *Synthetic Biology and Morality: Artificial Life and the Bounds of Nature*. Cambridge, MA: MIT Press, 2013.

Lentzos, F., Jefferson, C., and Marris, C. *Synthetic Biology and Bioweapons*. London: Imperial College Press, 2017.

Pahara, J., Dickie, C., and Jorgensen, E. *Hacking DNA with Rapid DNA Prototyping: Synthetic Biology for Everyone*. Sebastopol, CA: O'Reilly Press, 2017.

Sources of quotations

Cho, R. State of the Planet Blog, Columbia University Earth Institute. Available at: <http://blogs.ei.columbia.edu/2011/07/08/synthetic-biology-creating-new-forms-of-life/>, 2011.

Church, G. Interviewed for SynBioWatch. Available at: <http://www.synbiowatch.org/2012/10/how-synthetic-biology-will-change-us/>, 2014.

Kahn, J. 'Synthetic Hype: A Skeptical View of the Promise of Synthetic Biology', *Val. U. L. Rev.* 45/29. Available at: <http://scholar.valpo.edu/vulr/vol45/iss4/2>, 2011.

Kuiken, T. 'DIYbio: Low Risk, High Potential', *The Scientist*, 1 March. Available at: <https://www.the-scientist.com/?articles.view/articleNo/34443/title/DIYbio--Low-Risk--High-Potential/>, 2013.

Thomas, J. 'Synthia is Alive…and Breeding Panacea or Pandora's Box?', *ETC News Release* 20 May. Available at: <http://www.etcgroup.org/sites/www.etcgroup.org/files/publication/pdf_file/ETCVenterSynthiaMay202010.pdf>, 2010.

Willets, D. 'Statement Concerning the Establishment of BBSRC/EPSRC Synthetic Biology Research Centres in the UK'. Available at: <http://webarchive.nationalarchives.gov.uk/20140714082920/http://www.epsrc.ac.uk/newsevents/news/biologyresearchcentres/>, 2014.

Index